OCR
Electronics
For AS

Michael Brimicombe

CD-ROM INSIDE

Orders: please contact Bookpoint Ltd, 130 Milton Park, Abingdon, Oxon OX14 4SB. Telephone: (44) 01235 827720. Fax: (44) 01235 400454. Lines are open from 9.00 – 5.00, Monday to Saturday, with a 24 hour message answering service. You can also order through our website www.hoddereducation.co.uk

If you have any comments to make about this, or any of our other titles, please send them to educationenquiries@hodder.co.uk

British Library Cataloguing in Publication Data
A catalogue record for this title is available from the British Library

ISBN: 978 0 340 966 358

First Edition Published 2008
Impression number   10 9 8 7 6 5 4
Year   2012, 2011

Copyright © 2008 Michael Brimicombe

All rights reserved. No part of this publication may be reproduced or transmitted in any form or by any means, electronic or mechanical, including photocopy, recording, or any information storage and retrieval system, without permission in writing from the publisher or under licence from the Copyright Licensing Agency Limited. Further details of such licences (for reprographic reproduction) may be obtained from the Copyright Licensing Agency Limited, of Saffron House, 6–10 Kirby Street, London EC1N 8TS.

Hachette Livre UK's policy is to use papers that are natural, renewable and
recyclable products and made from wood grown in sustainable forests.
The logging and manufacturing processes are expected to conform to the
environmental regulations of the country of origin.

Typeset by Servis Filmsetting Ltd, Stockport, Cheshire
Printed and bound by CPI Group (UK) Ltd, Croydon, CR0 4YY for Hodder Education, an Hachette UK Company, 338 Euston Road, London NW1 3BH by CPI Group (UK) Ltd, Croydon, CR0 4YY.

Contents

Introduction	**v**
1 Simple digital systems	**1**
1.1 Digital inputs	1
1.2 Combining signals	5
1.3 Switching outputs	10
1.4 System diagrams	12
2 Digital from analogue	**19**
2.1 Resistive sensors	19
2.2 Op-amps and diodes	23
2.3 Delaying signals	29
3 Digital pulses	**37**
3.1 Single spikes	37
3.2 Oscillators	43
4 Logic systems	**49**
4.1 Truth tables	49
4.2 Logic system design	54
4.3 Only NAND gates	59
5 Storing signals	**67**
5.1 Bistables	67
5.2 Latches	72
5.3 Flip-flops	75
6 Negative feedback	**82**
6.1 Amplifiers	82
6.2 Voltage followers	86
6.3 Known gain	89
6.4 Summing signals	93
7 Counting pulses	**100**
7.1 Binary counters	100
7.2 Clocks	105
7.3 Continuous sequencers	110
7.4 One-shot sequencers	112

8 Amplifying audio — 120

    8.1 Audio systems — 120
    8.2 Filters — 127

9 Microcontrollers — 137

    9.1 Programmable systems — 137
    9.2 Hardware — 141
    9.3 Software — 145

Appendices — 159

    Formulae — 159
    Circuit symbols — 160
    Prefixes — 161
    Boolean algebra — 161
    Flowchart symbols — 161

Index — 162

Introduction

You cannot learn electronics by just studying a book. This is because electronics is about the use of physical devices to perform useful tasks in the real world. Like learning to walk or driving a car, you can only get it straight in your head by doing it.

So this book is only part of a course of study to prepare you for the OCR exams in AS Electronics. It introduces you to a small number of state-of-the-art components (such as integrated circuit logic gates, op-amps and microcontrollers), assuming that you have no previous experience of electronics other than basic electricity at the level of GCSE Science. As well as describing what these components do, you will be shown how to summarize their behaviour in algebraic and graphical form, using simple useful models of what are actually complex real devices. There will be no attempt to show how these devices can be assembled from simpler components, such as transistors, nor are there any explanations of the physics which makes these components possible. The OCR course aims instead to show how modern complex electronic systems, such as computers, editing suites and telephone systems can be put together from a limited number of basic systems which are available as discrete integrated circuits. Along the way, you will meet important techniques, such as block diagrams, Boolean algebra and negative feedback which have formed the theoretical backbone of the subject since its birth less than a century ago.

The other parts of the course can be found at www.hodderplus.co.uk/ocrelectronics. It contains

- exercises to consolidate your understanding
- full instructions for practical work
- data sheets on integrated circuits
- answers to the questions at the end of the chapters
- guidance for teachers and technicians.

The book has been written on the assumption that you will start at the beginning and work through to the end, section by section. You should do the practical work and exercises for each section before attempting the exam-style questions. You can only acquire the full flavour of the art which is modern electronics with the correct mixture of theory, practical and self-evaluation.

What does 'the expert choice' mean for you?

We work with more examiners and experts than any other publisher

- Because we work with more experts and examiners than any other publisher, the very latest curriculum requirements are built into this course and there is a perfect match between your course and the resources that you need to succeed. We make it easier for you to gain the skills and knowledge that you need for the best results.

- We have chosen the best team of experts – including the people that mark the exams – to give you the very best chance of success; look out for their advice throughout this book: this is content that you can trust.

More direct contact with teachers and students than any other publisher

- We talk with more than 100 000 students every year through our student conferences, run by Philip Allan Updates. We hear at first hand what you need to make a success of your A-level studies and build what we learn into every new course. Learn more about our conferences at **www.philipallan.co.uk**

- Our new materials are trialled in classrooms as we develop them, and the feedback built into every new book or resource that we publish. You can be part of that. If you have comments that you would like to make about this book, please email us at: **feedback@hodder.co.uk**

More collaboration with Subject Associations than any other publisher

- Subject Associations sit at the heart of education. We work closely with more Associations than any other publisher. This means that our resources support the most creative teaching and learning, using the skills of the best teachers in their field to create resources for you.

More opportunities for your teachers to stay ahead than with any other publisher

- Through our Philip Allan Updates Conferences, we offer teachers access to Continuing Professional Development. Our focused and practical conferences ensure that your teachers have access to the best presenters, teaching materials and training resources. Our presenters include experienced teachers, Chief and Principal Examiners, leading educationalists, authors and consultants. This course is built on all of this expertise.

CHAPTER 1

Simple digital systems

1.1 Digital inputs

Think about a computer keyboard.

Each key has one of just two states. It is either pressed or it is not pressed.

Each time that a key is pressed, a switch underneath it closes. A wire coming from the switch has its voltage set to either +5 V (plus five volts) or 0 V (zero volts). This **digital signal** passes along the wire to the rest of the computer, carrying information about the state of that key. Electronic circuits in the computer react to the signal, generating other digital signals . . ., but that's another story. Let's stay with the keyboard for the moment.

Switches

Look at the circuit diagram of Fig. 1.1.

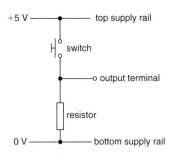

Fig 1.1 This circuit generates a +5 V signal at the output when the switch is pressed.

It shows two **components**, a **switch** and a **resistor**. Each component has two terminals, shown as lines coming from them. One terminal of the switch is connected to the **top supply rail**, a wire held at a voltage of +5 V by a power supply which is not shown in the diagram. Similarly, one terminal of the resistor is connected to the **bottom supply rail**, a wire held at 0 V by the same power supply. The **output terminal** comes from the connection between the switch and the resistor.

Fig. 1.2 shows the signal at the output terminal for each state of the switch.

Here is how it works. When the switch is closed, the top supply rail is connected directly to the output terminal. This means that both have to be at the same voltage of +5 V. The signal at the output is **high** or **1** (one). Opening the switch means that the output terminal is only connected to the bottom supply rail via the resistor. So the output terminal has to be at 0 V, giving a **low** signal or **0** (zero or nought).

Fig 1.2 The output signal depends on the state of the switch.

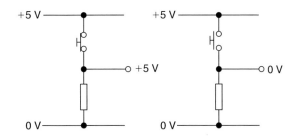

Charge

Let's take another look at what the resistor is doing when the switch is closed. One of its terminals is directly connected to +5 V, the other to 0 V. This **voltage drop** across the resistor allows **charge** to flow through it from the top rail to the bottom rail as shown in Fig. 1.3.

Fig 1.3 The arrows show the flow of charge through the resistor when the switch is closed.

The charge flows from the power supply into the top supply rail at +5 V. It then flows through the closed switch. The resistor and switch are connected **in series**, so all of the charge which leaves the switch flows into the resistor. Finally, the charge flows back to the power supply along the 0 V supply rail.

Electronic engineers don't worry too much about the exact nature of charge. They just accept that it is something which flows from high voltage to low voltage through metal wires, and that by adjusting its flow with electronic components they can accomplish the miracles of modern electronics. You should adopt the same attitude if you want to be successful in electronics.

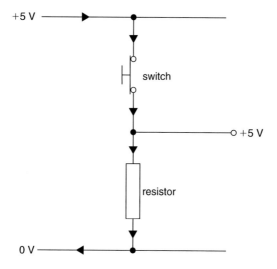

Meters

Charge flow in a resistor is measured by placing an **ammeter** in series with it as shown in Fig. 1.4.

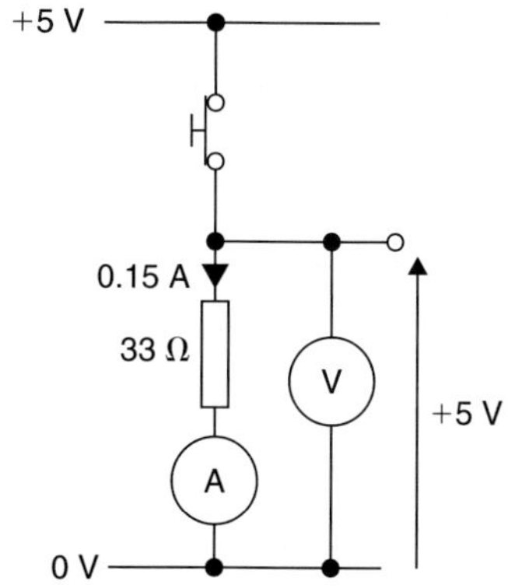

The ammeter measures the **current** in the resistor, a measure of how fast the charge is flowing through it. The direction of the current is shown by arrows on the wires of the circuit diagram. In this case it is downwards (from high voltage to low voltage) and it has a value of 0.15 A. The voltage drop across the resistor is measured by the **voltmeter** connected **in parallel** with it. Its reading is shown on the diagram as +5 V. The upwards arrow next to the meter points to the high voltage end.

Fig 1.4 The current in the 33 Ω resistor is 0.15 A when the voltage drop across it is 5 V.

Simple digital systems

Resistance

The current and voltage drop can be used to calculate the **resistance** of the resistor with this formula.

$$R = \frac{V}{I}$$

R is the resistance of the resistor, measured in **ohms** (Ω). It is a fixed quantity, depending on the construction of the resistor and is usually shown as a set of colour-coded rings painted on it. V is the **voltage drop** across the resistor, measured in **volts** (V) and always depends on the rest of the circuit. I is the **current** in the resistor, measured in **amperes** (amps) (A).

The ammeter reads 0.15 A when the switch is closed. What is the resistance of the resistor?

R = ?

V = 5 V

I = 0.15 A

$$R = \frac{V}{I} = \frac{5}{0.15} = 33 \, \Omega$$

Power

Current in a resistor always results in heating. This is because charge flowing from the power supply through a component transfers electrical energy to it. Resistors are designed to transfer this energy into heat. Engineers have to take account of this when designing circuits – a component which gets too hot is going to fail! The **power** of a component tells you how much it is being heated by the current in it. Power is calculated with this formula.

$$P = IV$$

P is the power of the component, measured in watts (W), V is the voltage drop across it in volts (V) and I is the current in it in amps (A).

So what is heating power of the 33 Ω resistor when the switch is closed?

P = ?

I = 0.15 A

V = 5 V

$$P = IV = 0.15 \times 5 = 0.75 \, W$$

Maximum rating

All resistors have a maximum power rating. If the actual power exceeds this value, the resistor will get too hot and may eventually burn out. The rating is fixed by the size of the resistor. The larger the rating, the easier it is for heat to escape. Here are some typical power ratings for resistors.

4 W   2 W   1 W   0.5 W   0.25 W

The actual power of the resistor in Fig. 1.4 is 0.75 W, so it should be rated at 1 W. A rating of 0.5 W will result in overheating, and 2 W will be unnecessarily bulky and expensive!

Pull-up resistors

Fig. 1.5 shows a different way of using a switch and resistor to generate a digital signal.

Fig 1.5 The resistor pulls the output high when the switch is not pressed.

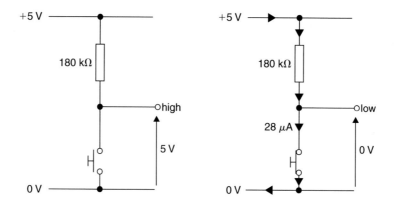

Consider the left-hand circuit. The switch is open, so no charge can flow through it. In fact there is no flow of charge anywhere, so the current in the resistor is 0 A. What is the voltage drop across it?

$V = ?$

$I = 0$ A

$R = 180$ k$\Omega = 180 \times 10^3$ Ω

$R = \dfrac{V}{I}$

$V = IR = 0 \times 180 \times 10^3 = 0$ V

(Notice the use the **prefix** k = $\times 10^3$ to deal with a very large resistance. You will find a list of all the prefixes you will need to use in the Appendix.)

If there is no voltage drop across the resistor, the output must be at the same voltage as the top supply rail. So the output is pulled high by the resistor, leaving a voltage drop of 5 V across the open switch.

Now look at the right-hand circuit. Charge flows through the closed switch as it pulls the output low, so there is a current in the circuit. The size of that current is fixed by the resistor.

$I = ?$

$V = +5$ V $- 0$ V $= 5$ V

$R = 180$ k$\Omega = 180 \times 10^3$ Ω

$R = \dfrac{V}{I}$

$I = \dfrac{V}{R} = \dfrac{5}{180 \times 10^3} = 2.8 \times 10^{-5}$ A $= 28$ μA

The resistor is rated at 125 mW. Will it overheat?

$P = ?$

$V = +5$V $- 0$ V $= 5$ V

$I = 28$ μA $= 28 \times 10^{-6}$ A

$P = IV = 28 \times 10^{-6} \times 5$

$= 1.4 \times 10^{-4}$ W $= 0.14$ mW

The actual power is much less than the rated power, so there is no risk of failure when the switch is closed.

Simple digital systems

1.2 Combining signals

Digital signals are altered and combined by electronic components called **logic gates**. These are produced as integrated circuits, and are available in several different types of technology. This book only tells you about the CMOS 4000 series of logic gates. Other technologies may be faster or handle larger powers, with different transfer characteristics, but CMOS gates are widely used, easily available and relatively foolproof for design purposes.

NOT

The simplest logic gate of all, the NOT gate, is shown in Fig. 1.6.

Fig 1.6 A NOT gate and its transfer characteristic.

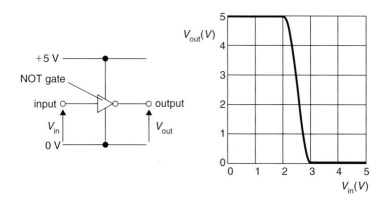

The circuit shows the NOT gate connected to supply rails at +5 V and 0 V. The graph shows how the voltage at the output terminal V_{out} is related to the voltage at the input terminal V_{in}. It represents the typical **transfer characteristic** of the device. All CMOS gates have the same transfer characteristic for values of V_{in} below +2 V and above +3 V, but there will be differences from one gate to another when V_{in} is between +2 V and +3 V. This does not really matter, as gates are designed to only handle signals whose voltage is +5 V or 0 V, well away from this region of uncertainty.

Truth table

The table below summarizes the behaviour of a CMOS NOT gate.

range of input voltages	state of output terminal
below +2 V	high
between +2 V and +3 V	uncertain
above +3 V	low

A	Q
0	1
1	0

Fig 1.7 The NOT truth table.

An alternative, neater representation, is shown in Fig. 1.7. It is known as a **truth table**.

The first column shows the two possible states of the input A. The second column shows the corresponding states of the output Q. Of course, 1 stands for a high voltage (above +3 V) and 0 stands for a low voltage (below +2 V).

By now, you will have realized why a NOT gate has its name. The input and output terminals do **not** have the same state as each other!

Output indicators

A switch and resistor can allow you to feed signals into a logic gate. But how can you detect the output signal? Light-emitting diodes (LEDs) are widely used to indicate the output states of logic gates. They convert electrical energy into light, giving a visual indication of the state of an output.

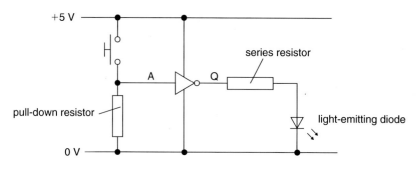

Fig 1.8 The LED indicates the state of the NOT gate's output terminal.

Take a look at Fig. 1.8. Notice that you cannot connect the LED directly to the output of the NOT gate. You need a resistor in series with it. This is because they have a typical voltage rating of only 2 V, hardly compatible with the 5 V from the output of the NOT gate.

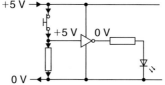

Fig. 1.9 shows what happens when the switch is pressed. The NOT gate input is connected directly to +5 V, so its output goes to 0 V. Since charge flows only from a higher voltage to a lower voltage, there will be no current in the LED and its series resistor. So the LED does not glow.

Fig 1.9 The LED does not glow when the switch is pressed.

Notice that although charge flows past the input of the NOT gate, all of it goes from the switch to its pull-down resistor. The current in the input terminal of a CMOS gate is tiny, far too small to worry about, so you might as well assume that there isn't any at all.

Series resistor

Fig. 1.10 shows the flow of charge when the switch is opened.

There is no current in the pull-down resistor, so it pulls the NOT gate input down to 0 V. The NOT gate output rises up to +5 V. Since there is only 2 V across the LED, the remaining $5 - 2 = 3$ V will be across the series resistor. This requires a current from the NOT gate output, typically about 1.5 mA. This comes from the top supply rail through one of the gate's supply connections. That current dumps energy into the LED, allowing it to glow with a power of 3 mW.

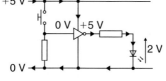

Fig 1.10 The LED glows when the switch is released.

The resistor value has to be carefully chosen to get the correct current in the LED.

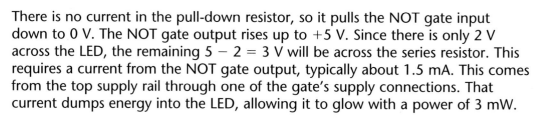

$R = ?$

$I = 1.5$ mA $= 1.5 \times 10^{-3}$ A

$V = 5 - 2 = 3$ V

$R = \dfrac{V}{I} = \dfrac{3}{1.5 \times 10^{-3}} = 2 \times 10^3 \, \Omega = 2 \, k\Omega$

Simple digital systems

LED bias

The transfer characteristic for an ideal LED is shown in Fig. 1.11.

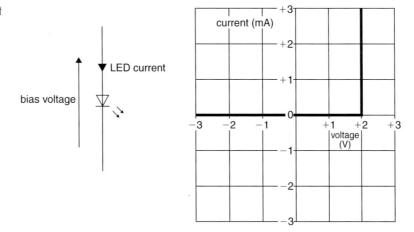

Fig 1.11 There is only current in an ideal LED when its bias voltage is +2 V.

The LED needs to have the correct **bias voltage** to glow. There will be no current in it until the bias voltage reaches +2 V. (The actual voltage required for a real LED to glow will be slightly different, depending on its colour.) However, once this value has been reached, the current rises very rapidly as the voltage is increased further. This is shown as a vertical line in the graph of Fig. 1.11; it is not as steep as this for real LEDs.

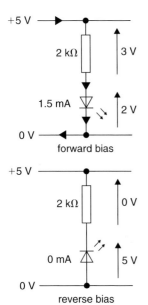

Polarity

LEDs, like most electronic components, have a **polarity**. They only work if they are correctly connected to their power supply. Take a look at Fig. 1.12. When the LED is **forward biased** it will glow. Charge can flow through it, with the current limited to 1.5 mA by the 2 kΩ series resistor. However, **reverse bias** results in no flow of charge at all (the bias voltage is negative) and the LED does not glow (even though the voltage drop across it is paradoxically 5 V).

Fig 1.12 The LED does not glow in reverse bias.

Indicating low

Although it may seem natural to use a glowing LED to indicate a high output, the opposite is often used. As Fig. 1.13 shows, the logic gate can be used to **sink** current from the LED and its resistor.

Note that the current of 28 μA that the switch **sources** into its 180 kΩ pull-down resistor is much smaller than the current of 1.5 mA that the NOT gate sinks from the LED. None of the charge which flows through the LED passes through the switch – it all comes from the top supply rail and enters the bottom supply rail through the gate's low voltage supply connection.

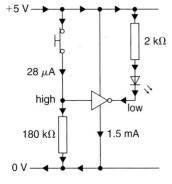

Fig 1.13 The LED glows each time the output goes low.

7

AND

The AND gate shown in Fig. 1.14 has two inputs (A, B) and a single output (Q).

Fig 1.14 An AND gate and its behaviour.

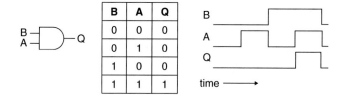

Each row of the truth table gives the state of the output Q for a different input state BA. Q is only 1 when the **binary word** BA is 11; otherwise Q is 0. In other words, Q is only high when B **and** A are high.

Fig. 1.14 also shows the **timing diagram** for an AND gate. It shows how the signal at Q changes as the signals at B and A change. AND gates are useful because the signal at one input can be used to control the flow through the gate of the signal at the other input. Use the timing diagram to convince yourself that when B is high, signals at A get through to Q, but when B is low Q is forced to stay low as well.

OR

OR gates can also be used to control the flow of signals. Look at the truth table of Fig. 1.15.

Fig 1.15 An OR gate and its behaviour.

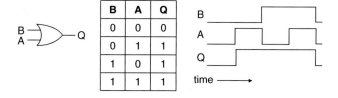

The timing diagram shows that the signal at A can only get through to Q when B is low. When B is high, Q is forced to stay high as well. In other words, Q is only high when A **or** B are high; otherwise Q is low.

NOR

It is no coincidence that the symbol for a NOR gate is a combination of the symbols for an OR gate and a NOT gate. Study the truth table of Fig. 1.16 and convince yourself that a NOR gate behaves like an OR gate followed by a NOT gate.

Fig 1.16 A NOR gate and its behaviour.

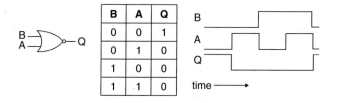

Simple digital systems

EOR

EOR stands for Exclusive-OR. As you can see from Fig.1.17, an EOR gate behaves like an OR gate, but only if one of the inputs is high. The EOR gate is sometimes known as a difference gate. This is because its output Q is only high when the two inputs B and A are different; otherwise Q is low.

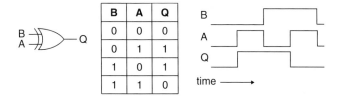

Fig 1.17 An EOR gate and its behaviour.

EOR gates are useful for selectively altering the state of a signal. Convince yourself that Q is A when B is low, but Q is the opposite of A when B is high.

NAND

The NAND gate is the most important of all the logic gates. Its symbol and truth table are shown in Fig. 1.18.

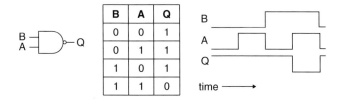

Fig 1.18 A NAND gate and its behaviour.

As its name and symbol suggest, a NAND gate behaves like an AND gate followed by a NOT gate. The output Q is only low when both of the inputs are high. So why does this make NAND gates so special? Well, as you will find out later, it turns out that all of the other two-input gates can be made from combinations of NAND gates, giving important savings in space on circuit boards. This is because you cannot get logic gates on their own; they are packaged in fours as shown in Fig. 1.19.

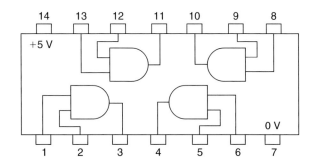

Fig 1.19 The arrangement of four AND gates in a single 4081 integrated circuit package.

OCR Electronics for AS

1.3 Switching outputs

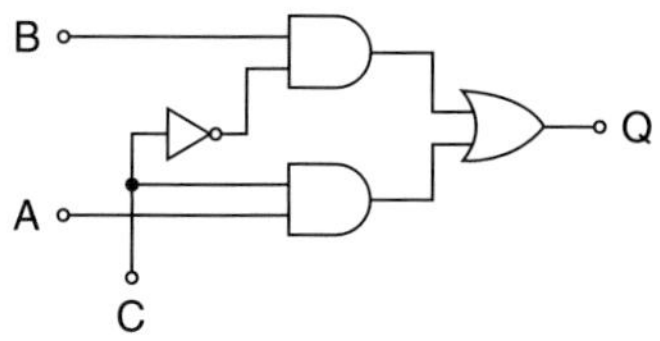

Logic gates are rarely used on their own. They are designed to take in digital signals, combine them according to their truth tables and then pass the combined signal onto other logic gates, as shown in Fig. 1.20. CMOS outputs are designed to feed their signals into CMOS inputs. Since those inputs need hardly any current, the outputs of logic gates are not good at either sinking or sourcing current. In fact, they struggle to deal with more than few milliamps. This is only a problem when the output signal of a logic gate has to control current-hungry components such as lamps, buzzers, heaters or motors. A logic gate needs a **driver** to interface with useful output devices.

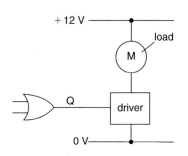

Fig 1.20 A logic system made from logic gates.

Drivers

Although there are several technologies available to make drivers, they all behave the same way. They have three terminals as shown in Fig. 1.21. The input terminal is connected to Q, the output of the logic system. A second terminal is connected to the 0 V supply rail. The third terminal sinks current from the **load** (a motor in this case) whenever Q goes high.

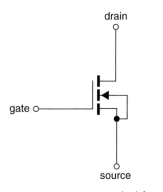

Fig 1.21 The driver allows the OR gate to control a motor.

MOSFETs

The acronym MOSFET stands for metal oxide semiconductor field effect transistor. It is the name given to one technology used to make electronic components. MOSFETs are used to make many devices (including CMOS logic gates), but are particularly effective at making drivers. Fig. 1.22 shows the circuit symbol of a suitable MOSFET. The voltage at the **gate** controls the flow of charge from the **drain** to the **source**, but hardly draws any current from the device controlling the gate.

The transfer characteristic of a typical MOSFET driver is shown in Fig. 1.23. When the voltage drop between gate and source V_{gs} is below 3 V, there is no current in the drain. No drain current means that the drain–source resistance R_{ds} is really big. It only starts to drop when V_{gs} reaches the MOSFET's **threshold voltage** – in this case it is 3 V, halfway between high and low for a logic gate. By the time V_{gs} has reached 5 V, the value of R_{ds} has dropped to a constant small value, 0.5 Ω in this case.

Fig 1.22 Circuit symbol for a MOSFET driver.

Fig 1.23 Transfer characteristic of a typical MOSFET.

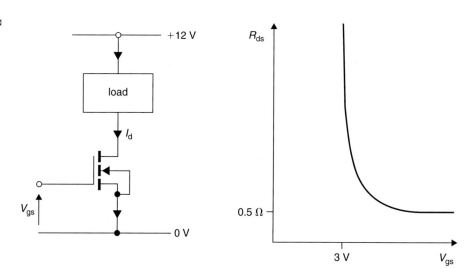

Simple digital systems

Maximum power

A MOSFET driver, like every other electronic component, has a maximum power rating. If its power exceeds this value, it will get too hot and fail. You need to be able to estimate the heating power of a MOSFET before you turn it on.

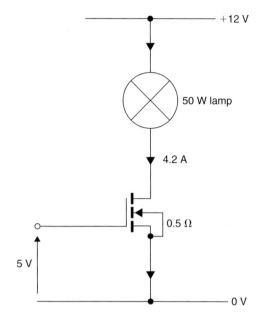

Fig 1.24 The MOSFET is rated at 20 W. Will it survive?

Study the circuit of Fig. 1.24. The MOSFET is sinking current from a lamp rated at 12 V, 50 W. It is rated at 20 W. Will it survive? The first step is to calculate the expected drain current.

$I = ?$

$V = 12$ V

$P = 50$ W

$P = IV$

$I = \dfrac{P}{V} = \dfrac{50}{12} = 4.2$ A

This current has to pass from drain to source, through a resistance of 0.5 Ω. This will require a voltage drop V_{ds} between drain and source.

$V = ?$

$I = 4.2$ A

$R = 0.5$ Ω

$R = \dfrac{V}{I}$

$V = IR = 4.2 \times 0.5 = 2.1$ V

Now you know the drain current and the drain–source voltage, you can calculate the heating power in the MOSFET.

$P = ?$

$V = 2.1$ V

$I = 4.2$ A

$P = IV = 4.2 \times 2.1 = 8.8$ W

Of course, this cannot be quite right, because if 12 V is dropped across the lamp and 2 V across the MOSFET, the top supply rail should be at +14 V. However, the calculation is good enough to indicate that the MOSFET is safe enough. Its heating power is at least half of the maximum rating.

OCR Electronics for AS

1.4 System diagrams

Electronics relies heavily on various different types of diagram to make complex systems easy to understand. This is because there is a limit to how much information you can hold in your head at once. As you will find out below, by progressively chunking components together into functional blocks, even the most daunting circuit can be reduced to a diagram which conveys its function with ease.

Circuit diagram

Consider the **circuit diagram** of Fig. 1.25.

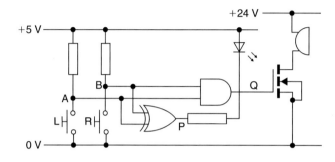

Fig 1.25 A complex circuit diagram.

The diagram shows the connections between all of the different components on the circuit. Each component has its own **symbol** (there is a list of them in the Appendix), and the wires connecting them together are shown as horizontal or vertical lines. Dots show where these wires are connected to each other. However, some connections are never shown. Power supply connections to integrated circuits, such as logic gates, are never shown as they would clutter up the diagram without adding any useful information. You have to assume that the logic gates are connected to the supply rails.

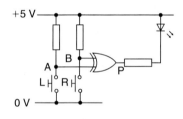

Fig 1.26

Block diagram

If you study Fig. 1.25 carefully, you will find that it can be separated into two distinct circuits. They are shown in Fig. 1.26 and Fig. 1.27.

Now we can focus on each circuit in turn. Let's start with Fig. 1.26, the simpler of the two, and represent it by a **block diagram** (Fig. 1.28).

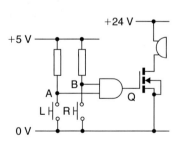

Fig 1.27

The block diagram groups circuit components together according to their function. There are two **input blocks**, each representing a switch and its pull-up resistor. Arrows from each input block show the flow of **information** into the **processor block** that represents the EOR gate. A single arrow going into the last block identifies it as the **output block**, representing the LED and its series resistor.

In a similar way, the circuit diagram in Fig. 1.27 can also be reduced to a block diagram. Notice that it has two processing blocks instead of just one (Fig. 1.29).

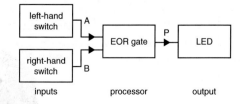

Fig 1.28 Block diagram of the circuit in Fig. 1.26.

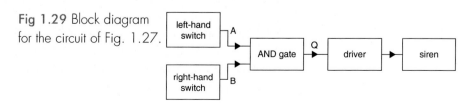

Fig 1.29 Block diagram for the circuit of Fig. 1.27.

Simple digital systems

Analysis

The block diagram of Fig.1.28 does not tell you at a glance what the system does. It just breaks it up into a small number of parts, each of which can be **analysed** on their own. Start by writing out truth tables for the input and output blocks, with a backward glance at their circuit diagrams.

switch L	signal A
open	high
closed	low

switch R	signal B
open	high
closed	low

signal P	LED
low	on
high	off

These can then be added on either side of the truth table for the EOR gate in the processing block.

switch R	switch L	B	A	P	LED
closed	closed	0	0	0	on
closed	open	0	1	1	off
open	closed	1	0	1	off
open	open	1	1	0	on

The columns for the internal signals can now be left out, leaving a truth table for the whole circuit of Fig. 1.26. This tells you that if the LED is off, only one of the switches is being pressed.

switch R	switch L	LED
closed	closed	on
closed	open	off
open	closed	off
open	open	on

Applying the same procedure to the block diagram of Fig. 1.29 gives you a truth table for Fig. 1.27. If the siren is on, then both switches have been released.

switch R	switch L	siren
closed	closed	off
closed	open	off
open	closed	off
open	open	on

Finally, the truth tables for the separate circuits can be combined to give a single table that tells you what the circuit of Fig. 1.25 does.

switch R	switch L	siren	LED
closed	closed	off	on
closed	open	off	off
open	closed	off	off
open	open	on	on

Imagine that the switches are built into a chair, one at the end of each arm. Sit in the chair and grip both arms, pressing the switches. This puts the LED on, but keeps the siren off. Releasing either arm turns the LED off, but the siren does not come on unless you release both arms.

OCR Electronics for AS

Synthesis

Block diagrams are also good starting points for designing circuits. You need to start with a **specification**, a sentence which explains what the whole system is supposed to do. Here's an example:

The heater only comes on when neither switch is pressed.

This can be translated into this truth table.

switch X	switch Y	heater
open	open	on
open	closed	off
closed	open	off
closed	closed	off

The block diagram is shown in Fig. 1.30. Two switches on the left, a logic gate (as yet unknown) to combine their signals and a heater as the output on the right.

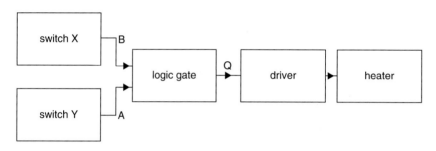

Fig 1.30 Block diagram which matches the specification.

Fig. 1.31 shows a suitable choice of **circuit elements** for the input and output blocks. (You could just as easily select switches with pull-up resistors instead of pull-down ones.)

Now add extra columns for the internal signals A, B and Q.

switch X	switch Y	B	A	Q	heater
open	open	0	0	1	on
open	closed	0	1	0	off
closed	open	1	0	0	off
closed	closed	1	1	0	off

Inspection of the columns for the internal signals suggests that the processor is a NOR gate. Joining the various circuit elements together gives the final circuit diagram shown in Fig. 1.32.

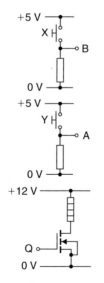

Fig 1.31 Circuit elements for the input and output blocks.

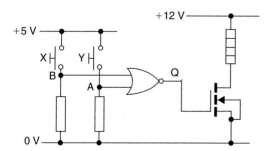

Fig 1.32 The final circuit diagram.

Simple digital systems

Questions

1.1 Digital inputs

1. Show how a switch and resistor can be arranged to create a signal of +5 V each time the switch is pressed, returning to 0 V when it is released. Explain the operation of the circuit, using the terms charge, current and voltage.

2. A 120 Ω resistor is connected directly across a pair of supply rails at +5 V and 0 V.

   (a) Calculate the current in the resistor.

   (b) Calculate the heating power of the resistor.

   (c) Select a suitable power rating for the resistor.

   Choose from this list. Justify your choice.

   500 mW   250 mW   125 mW

3. A switch and pull-up resistor are connected in series with a 5 V supply to generate a digital signal. An ammeter measures the current in the resistor and a voltmeter measures the voltage drop across the switch. Draw the circuit. Label all of the components.

4. An electronic system has supply rails that are 30 V apart. If all of the resistors are rated at a maximum power of 250 mW, what is the smallest safe value of resistor that can be used? Justify your choice with calculations.

5. This question is about the circuit shown in Fig. Q1.1.

   (a) Explain why there is no current in the resistor when the output is high.

   (b) Calculate the current in the resistor when the output is low.

   (c) State and explain how the state of the output terminal is determined by the switches L and R.

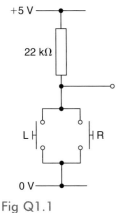

Fig Q1.1

1.2 Combining signals

1. Sketch a typical transfer characteristic for a NOT gate. Use it to explain what is meant by the symbols **1** and **0** in digital electronics.

2. The truth table for an Exclusive-Or gate is verified by connecting its input and output terminals to switches, resistors and an LED.

   (a) Draw a suitable circuit diagram.

   (b) Describe, in detail, how you would use the circuit to draw a truth table for the gate.

3. A particular LED is rated at 1.9 V, 2.2 mA.

   (a) Explain why a series resistor is required when an OR gate sources current into the LED.

   (b) Show how the LED and resistor should be connected to the OR gate.

   (c) Calculate a suitable resistance and power rating for the series resistor.

4. Here is the truth table for a logic gate.

   (a) Describe the behaviour of the logic gate.

   (b) Give the name and circuit symbol of the logic gate.

B	A	Q
0	1	1
1	1	0
1	0	1
0	0	1

15

OCR Electronics for AS

B	A	Q
0	0	
0	1	
1	1	

5. Complete this truth table for a NOR gate.

6. An LED can be in forward or reverse bias.

   (a) With the aid of circuit diagrams, explain what is meant by the terms **forward bias** and **reverse bias**.

   (b) Describe the electrical properties of an LED in forward and reverse bias.

1.3 Switching outputs

1. A MOSFET is a device with three terminals.

   (a) Draw the circuit symbol for a MOSFET and label the terminals.

   (b) Describe the electrical behaviour of a MOSFET.

2. Logic gates require a driver to pass on their signal to lamps, motors and heaters.

   (a) Explain why the driver is required.

   (b) Draw a circuit diagram to show a MOSFET as the driver between a NAND gate and a 9 V motor.

   (c) Describe and explain the signals required at the inputs of the NAND gate to make the motor shaft spin.

3. The graph in Fig. Q1.2 shows the transfer characteristic of a JCB58 MOSFET.

   (a) What is the threshold voltage for the JCB58?

   (b) In a particular circuit, a JCB58 sinks a current of 8.2 A from a load when the gate–source voltage is 4.5 V. Calculate the heating power of the MOSFET under these conditions.

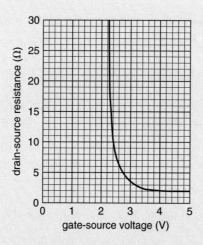

Fig Q1.2

1.4 System diagrams

1. Block diagrams and circuit diagrams are two different ways of representing electronic systems.

   (a) What are the differences between block and circuit diagrams?

   (b) What is the advantage of using a block diagram?

Simple digital systems

2 The circuit diagram in Fig. Q1.3 has two switches and an LED.

   (a) Draw a block diagram for the circuit.

   (b) Draw tables to show the behaviour of the two input blocks and the output block.

   (c) Describe the overall behaviour of the circuit.

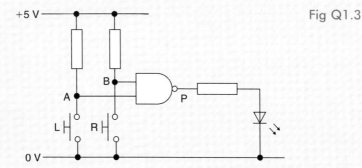

Fig Q1.3

3 A circuit is required to turn on a 12 V, 24 W lamp only when two switches are being pressed simultaneously.

   (a) Draw a block diagram for the system.

   (b) Explain the function of each block.

   (c) The switches use pull-up resistors to generate digital signals from a 5 V supply. Draw a circuit diagram for an input block, and draw up a table to show its behaviour.

   (d) Draw a circuit diagram for the whole system.

   (e) Explain the operation of the whole circuit. Include details of voltage and current at points within the circuit.

Learning summary

By the end of this chapter you should be able to:

- know that digital systems only process signals which are called 1 and 0
- know that digital signals are combined by logic gates
- use switches, resistors and LEDs to get signals in and out of logic gates
- know the truth tables for logic gates
- use MOSFETs to interface between logic gates and output devices
- use the rules $R = \dfrac{V}{I}$ and $P = IV$ for electronic devices
- use block diagrams to analyse the behaviour of digital systems

CHAPTER 2

Digital from analogue

2.1 Resistive sensors

Think about how you interact with a computer.

You can press keys on the keyboard or you can move the mouse around.

The mouse and the keyboard provide different types of signal to the the computer. Each key is either pressed or not pressed, so it makes a **digital signal**, with one of two values. The signal from the mouse depends on its position on the table, with many more than just two values. The position of the mouse is an **analogue signal**, with any value between limits rather than just two. Many useful signals are analogue, but computers can only deal with information in digital format. So there is a strong demand in electronics for circuits which can convert analogue signals into digital ones.

Thermistors

A **thermistor** is a device whose resistance depends on its temperature. Fig. 2.1 shows the transfer characteristic of a typical thermistor.

Fig 2.1 A thermistor's resistance falls as the temperature rises.

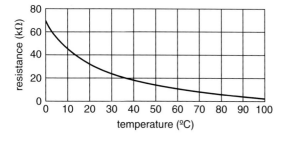

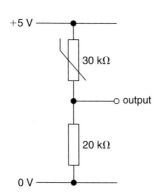

Fig 2.2 A thermistor in series with a fixed resistor makes a temperature sensor.

The value of a thermistor is often specified as its resistance at 25°C. So the thermistor featured in Fig 2.1 would be specified in a catalogue as 30 kΩ, dropping by a factor of ten for every 50°C increase of temperature. Fig. 2.2 shows how it might be made part of a temperature sensor circuit.

The arrangement of components is called a **voltage divider**. The two resistors in series (one fixed, the other variable) share the 5 V supply voltage between them, with the voltage at the output terminal dependent on the ratio of the two resistors.

OCR Electronics for AS

Voltage divider calculations

In order to calculate the voltage at the output of a voltage divider circuit, you need to remember three things:

- the total resistance is the sum of the two resistors.
- the current is the same in both resistors.
- the voltage drop across each is given by $V = IR$.

Take the situation in Fig. 2.3 as an example.

This is how to calculate the voltage of the output terminal. Firstly, work out the total resistance R between the two supply rails.

$R = ?$

$R_t = 30 \text{ k}\Omega$ $\qquad R = R_t + R_b = 30 \times 10^3 + 20 \times 10^3 = 50 \times 10^3 \ \Omega$

$R_b = 20 \text{ k}\Omega$

Secondly, calculate the current in that total resistance. You do this by replacing the pair of resistors with a single 50 kΩ resistor between the supply rails.

$I = ?$

$V = 5 - 0 = 5 \text{ V}$ $\qquad I = \dfrac{V}{R} = \dfrac{5}{50 \times 10^3} = 1.0 \times 10^{-4} \text{ A}$

$R = 50 \times 10^3 \ \Omega$

The third and final stage requires you to reinstall the two resistors in series. They have the same current as each other, 1.0×10^{-4} A or 100 μA. This is enough for you to calculate the voltage drop across just the 20 kΩ fixed resistor.

$V = ?$

$I = 1.0 \times 10^{-4} \text{ A}$ $\qquad V = IR = 1.0 \times 10^{-4} \times 20 \times 10^3 = 2.0 \text{ V}$

$R = 20 \times 10^3 \ \Omega$

So the output terminal sits 2 V above the 0 V supply rail.

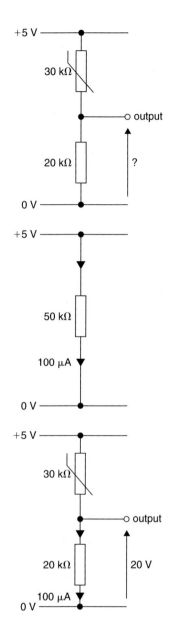

Fig 2.3 Three stages in calculating the output of a voltage divider.

Resistor ratios

If the bottom resistor of Fig. 2.3 has 2.0 V across it, then the top resistor must have a voltage drop of $5.0 - 2.0 = 3.0$ V across it. Notice that the ratio of the two resistors is the same as the ratio of their voltage drops:

$$\dfrac{30 \times 10^3 \ \Omega}{20 \times 10^3 \ \Omega} = \dfrac{3.0 \text{ V}}{2.0 \text{ V}}$$

This is always true provided that no current is drawn from the output terminal, thereby allowing both resistors to have the same current. Any current out of the output terminal will lower its voltage.

Digital from analogue

Sensor circuits

Thermistors are not the only component whose resistance varies with some aspect of their environment. For example, the resistance of an **LDR** (light-dependent resistor) depends on the intensity of light falling on it, falling rapidly as the light intensity increases. Fig. 2.4 shows how one might be made part of a voltage divider to make a light sensor.

Fig 2.4 An LDR configured to give a high signal when it is dark.

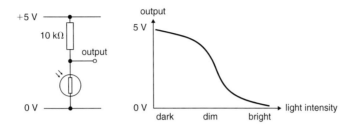

So how do you go about working out the transfer characteristic of this sensor?

Start off at one of the extremes of the environment. Let's go for dark. The resistance of the LDR will be large, much bigger than 10 kΩ. So most of the 5 V supply will be dropped across the LDR, resulting in a voltage close to the top supply rail at the output terminal. The output will be high.

Now consider the other extreme, where the light intensity is bright. The resistance of the LDR will be much lower than 10 kΩ, so it will drop very little of the 5 V supply compared with the fixed resistor. The voltage of the output terminal will be low, close to the bottom supply rail.

The region of maximum sensitivity, where the output varies most rapidly with light intensity, is fixed by the value of the fixed resistor. This happens when the LDR and the resistor have the same value: 10 kΩ in this case.

Position sensors

A **potentiometer** is a three-terminal device which can be used as a sensor. It has a shaft whose rotation fixes the resistance between its terminals. It is effectively a voltage divider, where the values of the two resistors depend on the setting of the shaft. Fig. 2.5 gives you some of the details.

Fig 2.5 The angle of the shaft sets the voltage at the output of this position sensor.

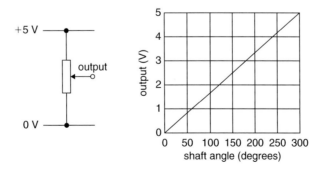

Here are the key properties of a potentiometer:

- The resistance between the **top** and **bottom** terminals remains constant.
- The resistance between the **wiper** and the other terminals depends on the angle of the shaft.

OCR Electronics for AS

Fig. 2.6 shows a voltage divider equivalent to a potentiometer. Rotating the potentiometer shaft increases one resistor but decreases the other, leaving the total resistance the same.

Fig 2.6 A potentiometer can be viewed as a pair of variable resistors in series.

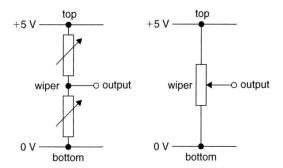

There are two types of potentiometer. **Rotary potentiometers** have a shaft that rotates the wiper, making it the natural choice for an angle sensor. **Linear potentiometers** have a wiper that moves in a straight line back and forth, allowing it to make a position sensor.

Setting signals

Suppose that you want to use a voltage divider to make a fixed voltage of +1.5 V from a pair of supply rails at +5 V and 0 V, as shown in Fig. 2.7. How do you go about it?

Start off by deciding how much current there should be in your voltage divider. The smaller it is, the less heat energy will be wasted in the resistors. However, it needs to be at least ten times larger than the current likely to the drawn from the output terminal, if the calculation is to work reasonably accurately. Let's suppose that a maximum current of 1 mA might be drawn from the output. This fixes the resistor current at 10 mA. Now you can calculate the total resistance of the circuit.

$R = ?$

$V = 5 - 0 = 5 \text{ V}$

$I = 10 \text{ mA} = 10 \times 10^{-3} \text{ A}$

$R = \dfrac{V}{I} = \dfrac{5}{10 \times 10^{-3}} = 500 \text{ }\Omega$

Now break the single 500 Ω resistor into two parts, in the ratio of the voltage drop required of them. The simplest way of doing this is to first consider the bottom resistor on its own.

$R = ?$

$V = 1.5 - 0 = 1.5 \text{ V}$

$I = 10 \times 10^{-3} \text{ A}$

$R = \dfrac{V}{I} = \dfrac{1.5}{10 \times 10^{-3}} = 150 \text{ }\Omega$

The top resistor is therefore 500 − 150 = 350 Ω.

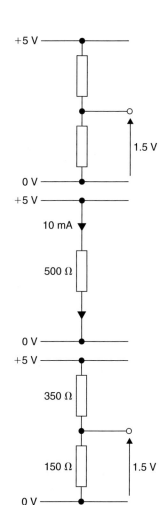

Fig 2.7 Three stages in selecting resistors for a voltage divider.

Digital from analogue

2.2 Op-amps and diodes

The light sensor shown in Fig. 2.8 could be useful for a camera. A high output from the voltage divider would indicate that there was enough light around to allow a picture to be taken, forcing the output of the NOT gate low. Unfortunately, the output of the NOT gate can be in the uncertain region between 1 and 0 if the light is just dim enough, so the output is still an analogue signal. You need a different type of processor to convert the analogue output of the voltage divider into a digital signal which is high or low, and never anything between. An operational amplifier (op-amp) does just this.

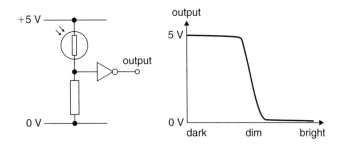

Fig 2.8 A bad way of converting an analogue signal into a digital one.

True digital

Fig. 2.9 shows how an op-amp, a diode and a pair of voltage dividers can make a light sensor whose output is always high or low, and never anything in between.

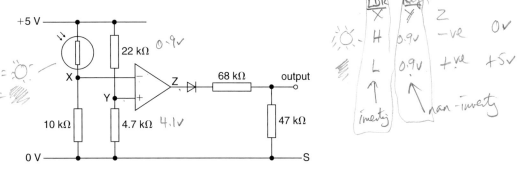

Fig 2.9 The op-amp converts the analogue light signal into an output which is digital – either high or low.

The circuit is quite complicated and contains a couple of electronic components which you haven't met yet: an op-amp and a diode. The block diagram of Fig. 2.10 breaks the circuit up into smaller blocks, making it easier to see what is going on.

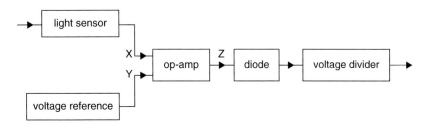

Fig 2.10 Block diagram for the circuit of Fig. 2.9.

At the left are two voltage dividers. One produces a fixed reference signal of +0.9 V. The output of the light sensor depends on the light intensity, being high in the light and low in the dark. These two signals enter the op-amp, which compares the two input signals making its output Z a positive voltage if the signal at Y is greater than the signal at X, otherwise it makes Z negative. The diode and voltage divider convert the two possible signals at Z into +5 V and 0 V.

OCR Electronics for AS

Op-amps

Like a logic gate, an op-amp has two input terminals and one output. Unlike a logic gate, it needs to be connected to supply rails at positive and negative voltages, so that its output can swing to a positive or a negative voltage. The circuit symbol is shown in Fig. 2.11.

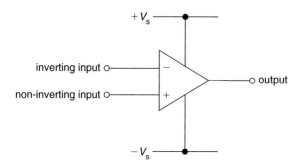

Fig 2.11 Inputs and outputs for an op-amp.

Op-amps are usually not fussy about their supply voltages. The popular TL084 op-amp works with supply rails from ±3 V to ±18 V. This book will assume from now on that all op-amps are run off supply rails at ±15 V and have the characteristics of a TL084. The two inputs are labelled **inverting** and **non-inverting**. These names reflect their effect on the output signal, shown in this table.

input condition	output signal
inverting above non-inverting	−13 V
inverting below non-inverting	+13 V

The output of an ideal op-amp **saturates** to within 2 V of its supply voltages of +15 V and −15 V, depending on which of its inputs is at the higher voltage. This behaviour is summarized in the graphs of Fig. 2.12.

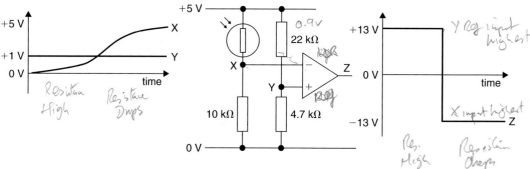

Fig 2.12 As the light intensity increases gradually, the output of the op-amp drops suddenly.

The output can change from one of these states to the other in about 2 μs, sourcing or sinking currents of up to 40 mA. The current at the inputs is much smaller than this, less than 1 nA, making it ideal for taking signals from voltage dividers with large resistors.

When $x > 0.9v$ $z = -13v$

When $x < 0.9v$ $z = +13v$

Digital from analogue

Diodes

Logic gates cannot handle negative voltages at their inputs, so the output of an op-amp cannot be fed directly into one. A **diode** is necessary to only let positive voltages through, as shown in Fig. 2.13.

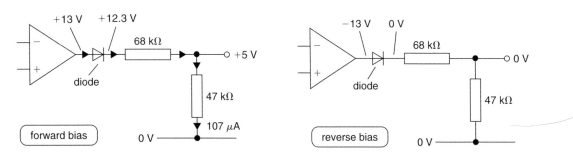

Fig 2.13 Charge only flows through the diode when it has forward bias.

A diode has two terminals, called the **anode** and the **cathode**. Charge can only flow from the anode to the cathode. For this to happen, the diode must be put in **forward bias**, with the anode at a higher voltage than the cathode. For a **silicon diode**, the voltage across a forward biased diode is about 0.7 V, regardless of the current in it (see the transfer characteristic of Fig. 2.14). So when the op-amp saturates at +13 V, the voltage drop across the pair of resistors is only 13 − 0.7 = 12.3 V. You can check for yourself that the current will be 107 μA, giving a +5 V signal at the output.

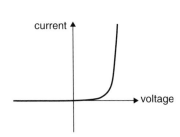

Fig 2.14 The current–voltage characteristic of a diode. Positive values of current and voltage correspond to forward bias.

The diode behaves quite differently when the op-amp saturates at −13 V. The op-amp is now trying to sink current instead of sourcing it, putting the diode in reverse bias. No charge can flow through it, so there is no current either. There will be no voltage drop across the resistors, so the output sits at 0 V.

The essential properties of an ideal silicon diode are summarized in this table. Although real diodes do not follow these rules exactly, they are good enough for most circuit designs.

	forward bias	reverse bias
voltage drop	+0.7 V	<+0.7 V
current	anything	nothing
resistance	small	large

Clamp diodes

There is more than one way of processing the output of an op-amp to make it compatible with a logic gate. The arrangement in Fig. 2.15 uses diodes to **clamp** the output voltage to within 0.7 V of a pair of supply rails.

Consider the situation where the op-amp has saturated positively. Charge flows through the 100 kΩ resistor and a diode to the +5 V supply rail, leaving the output at +5.7 V. Negative saturation allows the op-amp to sink current through the resistor and the other diode, leaving the output at −0.7 V. Although both of these voltages lie outside the digital range of 5 V to 0 V, they are not going to damage the inputs of logic gates.

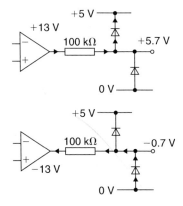

Fig 2.15 One diode is always in forward bias, clamping the output voltage close to the supply voltage.

OCR Electronics for AS

Range sensors

A single op-amp can only tell you if an analogue signal is above or below a given voltage. Two op-amps working together can tell you if an analogue signal is within a given range of voltages. For example, the circuit of Fig. 2.16 uses an LED to indicate that the output T of the temperature sensing voltage divider is between the voltages at U and L. This could be useful where the temperature of an object, such a baby's bath water, has to be between certain values.

Fig 2.16 *(left)* The LED only glows when T is between U and L.

Fig 2.17 *(right)* As the temperature increases, the LED comes on and then goes off again.

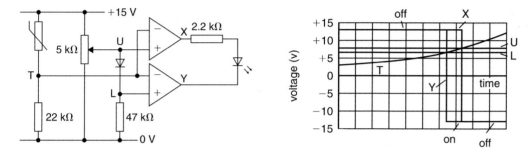

Each op-amp compares the thermistor signal T with one of the reference signals generated by the potentiometer, diode and fixed resistor. The upper reference signal U comes directly from the wiper of the potentiometer, so it can be varied from 0 V to +15 V. The lower reference signal L is 0.7 V lower than U, because of the forward-biased diode. The LED can only glow when it is forward biased. This requires X to be saturated at +13 V and Y at −13 V, a condition which is true only when T is between L and U. The voltage–time graph of Fig. 2.17 should make this clear.

Barchart sensors

Arrays of op-amps in parallel are widely used to make barchart displays. These indicate the size of their input signal with LEDs stacked side by side; the greater the signal, the more LEDs glow. The circuit of Fig. 2.18 shows how this can be done with an array of op-amps.

Fig 2.18 Barchart sensor, with a Zener diode to regulate the threshold voltages.

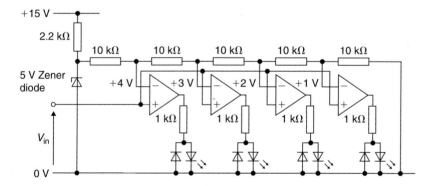

The **Zener diode** generates a fixed +5 V from the +15 V supply (for details see the next page). Five 10 kΩ resistors make a **resistor ladder** which generates voltages of +4 V, +3 V, +2 V and +1 V. Each of the four op-amps compares the incoming signal V_{in} with one of these four fixed voltages. If V_{in} is greater than the voltage at an non-inverting input, the output of an op-amp saturates at +13 V, putting the LED in forward bias and making it glow. The silicon diode in parallel with each LED clamps the reverse bias voltage to −0.7 V: well below the breakdown voltage of −5 V for a typical LED.

Digital from analogue

Breakdown voltage

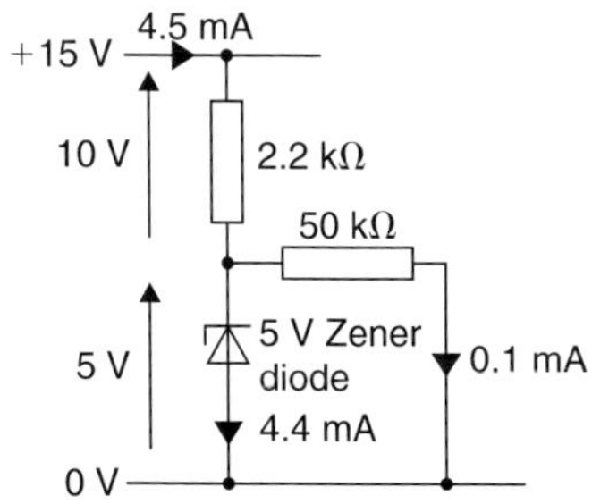

Fig. 2.19 takes a closer look at the Zener diode arrangement in the circuit of Fig. 2.18.

The Zener diode is in reverse bias. The current–voltage graph of Fig. 2.20 shows that in reverse bias there can be a current in a Zener diode provided that the voltage is greater than the **breakdown voltage**. Futhermore, the current increases sharply as the reverse bias voltage increases beyond the breakdown voltage. This means if you have enough reverse current, the reverse bias voltage is effectively independent of the current. For a BZX55C5V1 Zener diode, this point is reached when the current is 5 mA, with a reverse bias voltage of 5.1 V, although there is a possible 5 per cent variation between Zener diodes.

Fig 2.19 The Zener diode generates a fixed 5 V voltage.

Fig 2.20 Current–voltage characteristic for a Zener diode.

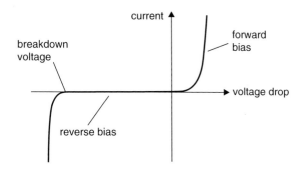

Zener current

You should never use a Zener diode without a series resistor to set the reverse current to a safe value. Consider Fig. 2.19. The current in the Zener diode comes from the +15 V supply rail via a 2.2 kΩ resistor. The voltage drop across the diode is 5 V, leaving 10 V across the resistor. This allows you to calculate the current in the resistor.

$I = ?$

$V = 15 - 5 = 10 \text{ V}$

$R = 2.2 \times 10^3 \text{ } \Omega$

$I = \dfrac{V}{R} = \dfrac{10}{2.2 \times 10^3} = 4.5 \times 10^{-3}$ A or 4.5 mA

Some of that current is diverted into the resistor ladder of five 10 kΩ resistors in series.

$I = ?$

$V = 5 - 0 = 5 \text{ V}$

$R = 5 \times 10 \text{ k}\Omega = 50 \times 10^3 \text{ } \Omega$

$I = \dfrac{V}{R} = \dfrac{5}{50 \times 10^3} = 1.0 \times 10^{-4}$ A or 0.1 mA

This leaves 4.5 − 0.1 = 4.4 mA for the Zener diode, close enough to the rated current of 5 mA. The power rating of the Zener diode is 500 mW, so it is in no danger of overheating with this current.

$P = ?$

$V = 5 \text{ V}$

$I = 4.4 \times 10^{-3} \text{ A}$

$P = IV = 4.4 \times 10^{-3} \times 5$

$= 2.2 \times 10^{-2}$ W or 22 mW

OCR Electronics for AS

Stable signals

Fig. 2.21 shows two different ways of obtaining the same reference signal for an op-amp. One uses a voltage divider, the other a Zener diode. Which should you use?

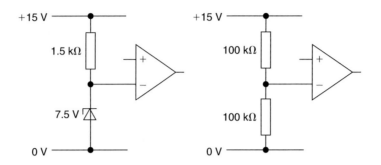

Fig 2.21 Two ways of generating a +7.5 V reference signal for an op-amp.

The circuit on the left uses a Zener diode. This means that the voltage at the inverting terminal will hardly change should the supply voltage change. In other words, you do not have to rely on the top supply rail being at exactly +15 V to obtain the reference signal. Although moving the supply rail up to +18 V or down to +12 V will considerably alter the current in the Zener diode, the steepness of the current–voltage characteristic means that the voltage across the diode will hardly change at all. This would not be the case for the circuit on the right. Moving the supply up or down by 3 V results in a massive 1.5 V change in the reference voltage. On the other hand, it draws much less current than the Zener diode (75 μA instead of 5 mA), thereby saving energy.

Digital from analogue

2.3 Delaying signals

Digital signals are rarely static. They usually change as time goes on, changing the state of any logic gates that they go into. Logic gates respond to changes in their input conditions very quickly – in about 250 ns for a CMOS gate. There are times when you need to slow down the response of a logic system, to delay any change of output when the input conditions change. This requires the use of capacitors, components which behave like small rechargeable batteries.

Shifting edges

The timing diagram in Fig. 2.22 shows how the output of a NOT gate reacts almost instantly to changes at its input. There is no current in the 47 kΩ resistor, so any change of voltage at the left-hand terminal is immediately transferred to the right-hand end.

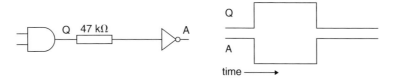

Fig 2.22 A changes instantly when Q changes.

The situation changes when a **capacitor** is connected between the input of the NOT gate and the 0 V supply rail, as shown in Fig. 2.23. Each time that Q changes state, there is a time delay of 1 ms before the equivalent change of state at A.

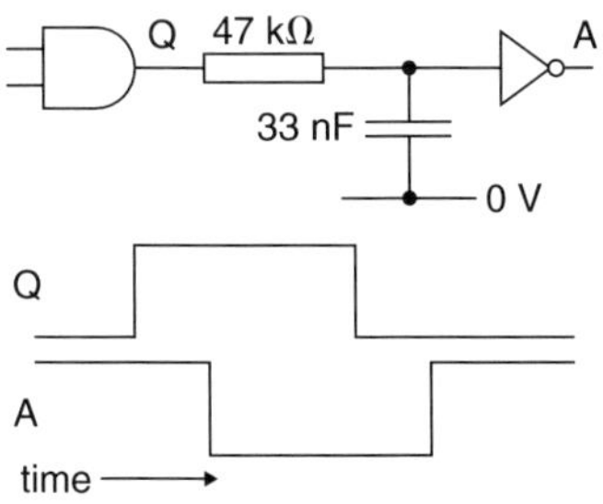

Changing charge

So how does a capacitor accomplish this trick of being able to tell the time? It cannot manage it on its own. The resistor in series with it is just as important in determining the effect of the capacitor.

Study the two scenarios presented in Fig. 2.24.

On the top, Q is raised to +5 V. The AND gate sources current into the resistor, allowing charge to flow onto the **top plate** of the capacitor. At the same time, charge flows off the **bottom plate** into the 0 V supply rail. Charge cannot actually flow right through a capacitor, as implied by the gap in its symbol, so it builds up on the top plate as time goes on. As the charge builds up the voltage V of the top plate increases. It reaches the CMOS threshold voltage of +2.5 V after 1 ms and the output of the NOT gate changes in 250 ns.

After a few milliseconds, V reaches 5 V and the flow of charge stops. The capacitor is **charged** and there is no longer any current in the resistor.

On the bottom, Q has been lowered back to 0 V. The AND gate sinks current from the resistor, allowing the charge on the top plate to flow through the resistor into Q. At the same time, charge flows onto the bottom plate from the 0 V supply rail, and the voltage V of the top plate drops as time goes on. A few milliseconds later V reaches 0 V and the capacitor is **discharged**.

Fig 2.23 The capacitor delays the rising and falling edges of the output signal.

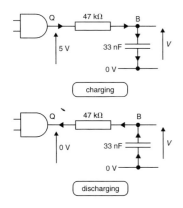

Fig 2.24 Charging and discharging the capacitor through the resistor.

Charging

Fig. 2.25 shows how the voltage of the top plate of a 33 nF capacitor changes with time as it is charged up from a 5 V signal through a 47 kΩ resistor.

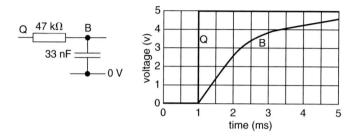

Fig 2.25 The capacitor charges slowly when Q goes from low to high.

Notice how the voltage at B does not rise steadily. It rises most rapidly at the start and rises less and less rapidly as time goes on, aiming towards, but never quite reaching, the voltage at Q. This type of change is known as an **exponential**. The time scale of the exponential is determined by this formula.

$$\tau = RC$$

The **time constant** τ of the circuit is measured in seconds (s). R is the resistance measured in ohms (Ω) and C is the **capacitance** measured in **farads** (F). For the circuit in Fig. 2.25, the time constant is 1.6 ms.

$\tau = ?$

$R = 47$ kΩ $\qquad \tau = RC = 47 \times 10^3 \times 33 \times 10^{-9}$

$C = 33$ nF $\qquad\qquad\; = 1.6 \times 10^{-3}$ s or 1.6 ms

The time constant is **not** how long it takes for the capacitor to charge up. In one time constant, the capacitor is at only 63 per cent of its eventual full charge. It reaches 95 per cent of its full charge after three time constants. The most useful way of using the time constant τ is given by this formula.

$$T_{1/2} = 0.7\tau$$

$T_{1/2}$ is how long it takes for the capacitor to reach 50 per cent of its full charge. Let's calculate this for the circuit of Fig. 2.25.

$T_{1/2} = ?$ $\qquad T_{1/2} = 0.7\tau = 0.7 \times 1.6 \times 10^{-3}$

$\tau = 1.6$ ms $\qquad\qquad\; = 1.1 \times 10^{-3}$ s or 1 ms

So the top plate of the capacitor reaches 2.5 V about 1 ms after Q shoots up to 5 V. Fig. 2.26 shows what effect this has on the NOT gate connected to the top plate; 2.5 V is the point at which A changes from high to low.

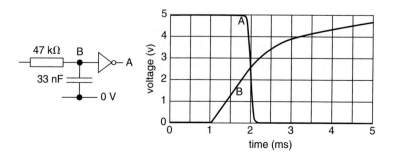

Fig 2.26 The output of the logic gate changes state as its input sweeps past 2.5 V.

Digital from analogue

Discharging

The graph of Fig. 2.27 shows what happens to the voltage of the top plate when Q is subsequently returned low. As with charging, the change in voltage is exponential. It drops most rapidly at the start, dropping less and less rapidly as time goes on, aiming towards, but never quite reaching, the new voltage at Q. The time taken for the capacitor to lose half of its initial charge is given by this formula.

$$T_{1/2} = 0.7\tau$$

As you can see from the graph, V drops from 5 V to 2.5 V in about 1 ms.

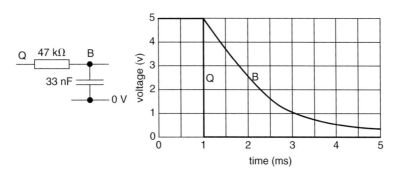

Fig 2.27 The capacitor discharges slowly when Q goes from high to low.

Switches and capacitors

The ability of a capacitor to store charge gives it the property of memory. Indeed, capacitors are at the heart of modern computer memory systems. Fig. 2.28 shows how a capacitor can be used to remember, for a short while, that a switch has been recently pressed.

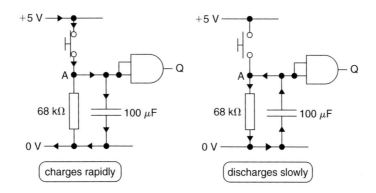

Fig 2.28 The capacitor charges rapidly but discharges slowly.

Pressing the switch charges up the capacitor instantly. The resistance of the charging circuit is very low, being just a closed switch, so there is a very quick burst of current in the supply rails. The top plate of the capacitor reaches +5 V straight away. When the switch is opened, the capacitor can only discharge through the 68 kΩ resistor in parallel with it. This takes time, so the voltage at A drops slowly.

Time delay

Suppose that you press the switch of Fig. 2.29 and then release it. How long do you have to wait before Q goes low?

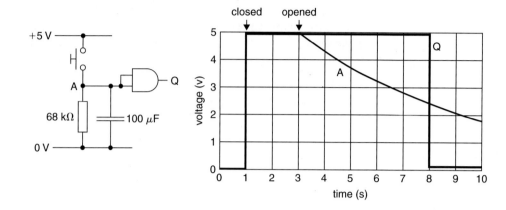

Fig 2.29 Q falls when A reaches 2.5 V.

The time taken for the voltage of the top plate to reach the critical threshold of 2.5 V is calculated in two steps.

$\tau = ?$

$R = 68$ kΩ $\qquad\qquad \tau = RC = 68 \times 10^3 \times 100 \times 10^{-6} = 6.8$ s

$C = 100$ μF

The time constant is then used to calculate the time taken for 50 per cent discharge.

$T_{1/2} = ?$

$\tau = 6.8$ s $\qquad\qquad T_{1/2} = 0.7\tau = 0.7 \times 6.8 = 4.8$ s (or 5 s)

So five seconds after the switch is released, Q drops low again.

Digital from analogue

Questions

2.1 Resistive sensors

1. A sensor circuit contains a thermistor and 12 kΩ resistor connected across supply rails at +9 V and 0 V.

   (a) Describe the electrical properties of a thermistor.

   (b) The voltage at the output of the sensor must increase with increasing temperature. Draw a circuit diagram for the sensor.

   (c) Calculate the voltage at the output of the sensor when the thermistor has a resistance of 7.2 kΩ.

2. A pair of resistors are used to generate a signal at +2 V from a pair of supply rails at +6 V and 0 V. The current in the output terminal is less than 50 µA. Calculate suitable values for the resistors and draw a circuit diagram for their arrangement.

3. This question is about the light sensor circuit shown in Fig. Q2.1.

   (a) Describe the electrical properties of both components.

   (b) Describe and explain how the signal at the output is determined by the lighting conditions of the circuit.

4. A 50 kΩ potentiometer is used to generate a variable voltage from +5 V to 0 V at one input of an AND gate, and a pair of resistors give a fixed +2.5 V at the other input.

   (a) Draw a circuit diagram for the arrangement.

   (b) The current in the resistors is required to be one-tenth of that in the potentiometer. Calculate the value of each resistor.

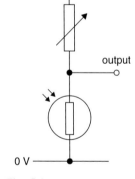

Fig Q2.1

2.2 Op-amps and diodes

Assume that the outputs of all the op-amps in these questions saturate at ±13 V.

1. The circuit in Fig. Q2.2 is a temperature sensor.

   (a) Describe the behaviour of the op-amp in the circuit.

   (b) Do a calculation to show that the voltage at B is about +4 V.

   (c) Explain why the signal at the output is 0 V when the thermistor is hot.

   (d) Calculate component values for the unlabelled resistors which hold the output terminal at +4 V when the thermistor is cold. The current at the output terminal will be less than 10 µA.

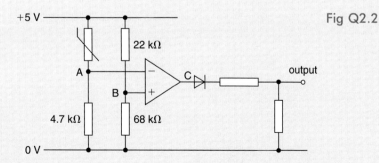

Fig Q2.2

2 The circuit in Fig. Q2.2 (on the previous page) contains a diode.

  (a) Sketch a current–voltage graph for a diode, indicating the regions of forward and reverse bias.

  (b) Describe the different electrical behaviour of a diode in forward and reverse bias, using the words current, resistance and voltage.

  (c) Show how a resistor and a pair of diodes can be used to clamp the output of an op-amp so that it can be safely connected to a logic gate.

3 The circuit in Fig. Q2.3 contains an LED rated at 2 V, 10 mA.

  (a) Calculate a suitable value for the current-limiting resistor in series with the LED.

  (b) Do calculations to show that U and Z are at 10 V and 7.5 V, respectively.

  (c) The voltage at W is set to 0 V and slowly increased to +15 V. Draw voltage–time graphs, on the same axes, for the voltages at W, X and Y during this time.

  (d) Explain why the LED only glows when the voltage at W is between 10 V and 7.5 V.

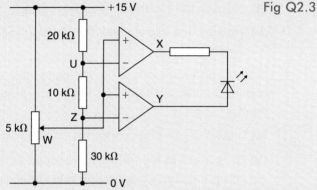

Fig Q2.3

4 A Zener diode is connected in series with a 220 Ω resistor between supply rails at 15 V and 0 V. The diode, which has a power rating of 500 mW, is connected in reverse bias to generate a signal at 5.1 V.

  (a) Draw a circuit diagram of the arrangement.

  (b) Sketch a current–voltage graph for the Zener diode, indicating the regions of forward and reverse bias. Mark important values on the voltage axis.

  (c) Calculate the current in the Zener diode and use it to show that the diode will not overheat in the circuit.

  (d) Give a reason why using a Zener diode to generate a fixed voltage signal is better than using a voltage divider.

Digital from analogue

2.3 Delaying signals

Assume that the logic gates in these questions have threshold voltages of +2.5 V.

1. The circuit of Fig. Q2.4 contains an RC network between a pair of NOT gates.

   (a) Calculate the time constant of the RC network.

   (b) The voltage at A suddenly changes from +5 V to 0 V. Sketch voltage–time graphs for the signals at X, Y and Q when this happens. Add scales to both axes.

   (c) Explain why the signal at Q is the same as the signal at A, but delayed by 15 ms.

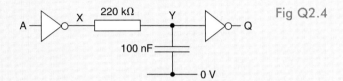

Fig Q2.4

2. A circuit is required to behave as follows:
   - An LED glows as soon as a switch is pressed.
   - The LED suddenly stops glowing ten seconds after the switch is released.

   (a) Draw a circuit diagram to show how the circuit can be built from a pair of resistors, a capacitor, a switch, an LED and a NOT gate.

   (b) Explain how the circuit operates.

   (c) Select component values to give the required time delay. Use calculations to justify your choice.

3. Sketch voltage–time graphs, with scales on both axes, to show the following signals.

   (a) The voltage across a 100 μF capacitor being charged up from a 12 V supply through a 47 kΩ resistor.

   (b) The voltage across a 22 nF capacitor, initially charged to 5 V, being discharged through a 33 kΩ resistor.

Learning summary

By the end of this chapter you should be able to:

- know the transfer characteristics of LDRs, thermistors and potentiometers
- calculate the output of a voltage divider circuit
- select resistor values of a voltage divider circuit for a given output
- know the difference between analogue and digital signals
- know the transfer characteristic of an op-amp
- know the transfer characteristics of diodes
- know how to use capacitors to delay digital signals

CHAPTER 3

Digital pulses

3.1 Single spikes

By now, you should be happy with the idea that electronic signals carry information. So the signal at A in Fig. 3.1 carries information about the state of the switch to the NOT gate. A high signal codes for a closed switch and a low signal codes for an open one.

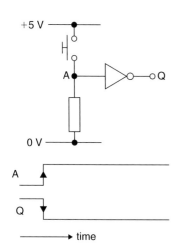

Fig 3.1 Pressing the switch feeds a rising edge into the NOT gate.

The timing diagram shows how the information changes with time as the switch is closed. Pressing the switch suddenly changes the voltage at A from 0 V to +5 V, feeding a **rising edge** into the NOT gate. The NOT gate responds by feeding out a **falling edge**. The timing of those **edge signals** (shown by rising and falling arrows in the timing diagram) contains information about when the switch is pressed or released. In fact, edge signals code for not only the closing and opening of the switch, they also tell you precisely when these events happen. The timing of a signal is often an important component of the information it contains.

Pulses

Suppose that you are interested **only** in **when** the switch was pressed? Fig 3.2 shows how the insertion of an RC network into the circuit codes the act of pressing the switch into a **pulse**.

Notice that the edge at the start of the pulse contains all of the useful information about the switch. It not only tells you that the switch has been closed (shown by the falling edge), but also when it happened. There are many situations where this is all you are interested in knowing – think about general knowledge quizzes. RC networks allow you to make brief signals which only contain information about the timing of events.

Fig 3.2 Pressing the switch pulses the output of the NOT gate low.

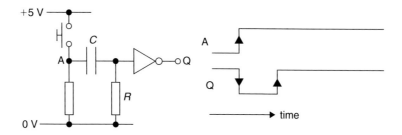

Spikes

Converting edges into pulses is a three-stage process:

1. An RC network converts the edges of a signal into **spikes** (short-lived signals above or below 0 V).
2. A clamp diode suppresses any negative spikes.
3. A logic gate converts positive spikes into pulses with clean rising and falling edges.

Let's start with the effect of the RC network as shown in Fig. 3.3.

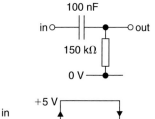

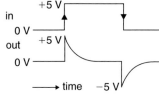

Fig 3.3 The RC network converts rising and falling edges into positive and negative spikes.

So what is going on? To start with, the output of the network is 0 V. Suddenly raising the left-hand plate of the 100 nF capacitor to +5 V causes it to start charging up. Charge immediately flows onto the left-hand plate from the source of the rising edge. At the same time, charge flows off the right-hand plate through the 150 kΩ resistor, until its voltage reaches 0 V. This does not take long.

$\tau = ?$

$R = 150$ kΩ

$C = 100$ nF

$\tau = RC = 150 \times 10^3 \times 100 \times 10^{-9}$

$= 1.5 \times 10^{-2}$ s or 15 ms

As you can see from Fig. 3.4, the whole process is almost over in a couple of time constants. The current in the resistor falls to zero and the output voltage drops back to 0 V almost immediately.

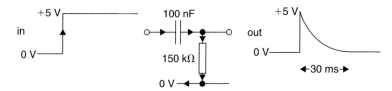

Fig 3.4 A rising edge charges up the capacitor in about 30 ms.

A subsequent falling edge at the input allows the capacitor to discharge (Fig. 3.5). The arrival of the edge suddenly forces the left-hand plate back to 0 V. Charge starts to flow off the left-hand plate into the source of the falling edge, and at the same time charge flows onto the right-hand plate through the resistor. As before, after a couple of time constants ($2 \times 15 = 30$ ms), most of the action is over, leaving both plates at 0 V.

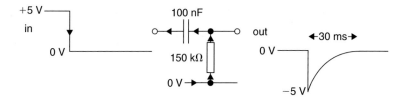

Fig 3.5 A subsequent falling edge discharges the capacitor, pulling the output below 0 V for a short time.

The operation of an RC network connected like this as a **spike generator** can be summarized as follows:

- Rising edges result in positive spikes.
- Falling edges result in negative spikes.
- The spikes only last for a couple of time constants.

Digital pulses

Spike suppression

Notice how the flow of charge through the resistor in Fig. 3.5 forces the right-hand plate to a negative voltage. This **negative spike** has a voltage which falls outside the permitted range of +5 V to 0 V for a digital signal, potentially causing problems for any logic gate it might enter. However, it is easily suppressed by connecting a diode in parallel with the output as shown in Fig. 3.6.

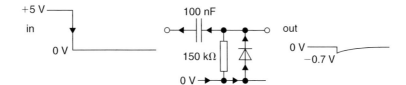

Fig 3.6 Negative spikes put the diode into forward bias, quickly discharging the capacitor

As soon as the output goes below −0.7 V, the diode becomes forward biased. Most of the current which is discharging the capacitor now flows through the diode instead of the resistor. This has two consequences:

- The output voltage cannot go below −0.7 V.
- The time constant becomes much smaller.

A forward-biased diode has a relatively small resistance, so it discharges the capacitor quickly. The diode not only **clamps** the output of the spike generator to −0.7 V, it shortens the length of the negative spike. This is illustrated in the timing diagram in Fig. 3.7.

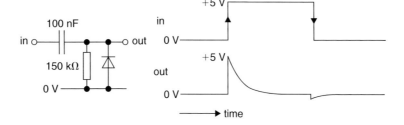

Fig 3.7 The clamp diode in a spike generator suppresses the negative spikes from falling edges.

Input protection

You do not actually need to include clamp diodes in logic circuits because they are already there. This is because CMOS logic gates are equipped with clamp diodes at each of their inputs. These protect the inputs from voltages more than 0.7 V above and below the supply rail voltages. Without these diodes, logic gates would be easily damaged by static electricity. The arrangement of diodes is shown in Fig. 3.8.

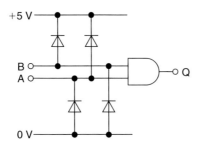

Fig 3.8 Clamp diodes at the inputs of logic gates.

Spike-to-pulse conversion

Placing a NOT gate on either side of an RC network makes a useful system for converting a falling edge into a low-going pulse. The arrangement is shown in Fig. 3.9.

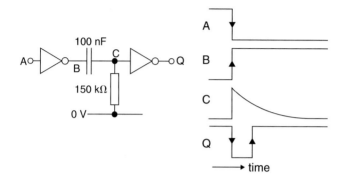

Fig 3.9 An edge-to-pulse converter. A falling edge at the input makes the output go low for a short time.

The timing diagrams show what happens when a falling edge arrives at the input A. The signal at B shoots straight up to +5 V, dragging the signal at C up at the same time, forcing the signal at Q to fall immediately to 0 V. As the capacitor charges up through the resistor, the voltage at C drops exponentially from +5 V to 0 V. As the signal at C drops past the threshold voltage of 2.5 V, the output of the right-hand NOT gate changes from 0 V to +5 V, terminating the pulse.

Pulse width

The length of time for which Q stays low can be calculated with this formula.

$$T = 0.7RC$$

The **duration** T of the pulse is measured in seconds (s). R is the resistance measured in ohms (Ω) and C is the capacitance measured in farads (F). The factor of 0.7 arises from the fact that the threshold voltage for CMOS gates is half of their supply voltage.

So how long does Q stay low when a falling edge arrives at A?

$T = ?$

$R = 150\ k\Omega$ $\qquad\qquad T = 0.7RC = 0.7 \times 150 \times 10^3 \times 100 \times 10^{-9}$

$C = 100\ nF$ $\qquad\qquad\qquad = 1.0 \times 10^{-2}$ or 10 ms

This value is just an estimate. In practice, the threshold voltage of a NOT gate varies from one integrated circuit to another, so actual the actual pulse width could be anywhere between 12 ms and 8 ms.

Premature end

The circuit of Fig. 3.9 has one major defect. The duration of the pulse at Q is only given by $T = 0.7RC$ if the signal at A remains low until the pulse has finished. Returning A back to +5 V before the end of the output pulse, brings it to an end immediately as shown in Fig. 3.10. As soon as A goes high, B goes low, forcing C to drop below 0 V. The clamp diode at the input of the NOT gate discharges the capacitor quickly, bringing the output pulse to a premature end.

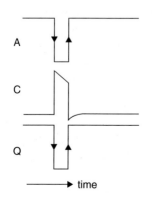

Fig 3.10 The resulting output when the duration of the input signal is too short.

Digital pulses

Monostable action

Replacing the NOT gates with NAND gates solves the problem as shown in Fig. 3.11. The circuit is called a **monostable**. It has the following properties:

- A falling edge at T triggers the appearance of a low-going pulse at Q.
- The duration of the pulse is 0.7RC.
- Once the pulse has started, it cannot be turned off by signals at T.

Fig 3.11 The duration of the pulse from a monostable does not depend on the duration of the triggering pulse.

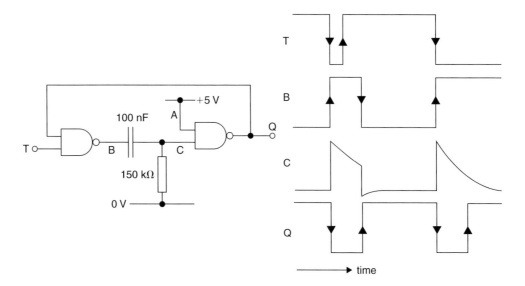

So how does the monostable of Fig. 3.11 manage to solve the shortcomings of the edge-to-pulse converter of Fig. 3.9? The secret lies in the **feedback loop** provided by the connection of the output Q to the left-hand NAND gate at the input. While Q is low, the output of that gate has to be high, regardless of the state of the input T. So once the falling edge at T has pulled Q low, the signal at B has to stay high until both Q and T have gone high again. The right-hand NAND gate acts as a NOT gate, ensuring that the pulse ends when C has dropped to half of its initial value.

Resistor values

You cannot use any value of resistor in a monostable. This is because the charge required to charge and discharge the capacitor has to flow through the output of the left-hand NAND gate. CMOS logic gates cannot supply much more than a couple of milliamps of current at their output, so a you need to ensure that the charging current never exceeds this limit. A choice of at least 10 kΩ will always work as the maximum current will only be 0.5 mA.

$I = ?$

$V = 5$ V

$R = 10$ kΩ

$$I = \frac{V}{R} = \frac{5}{10 \times 10^3} = 5 \times 10^{-4} \text{ A or } 0.5 \text{ mA}$$

The current in the input of a CMOS logic gate is tiny, so you do not have to worry about the maximum size of resistor. You can go all the way up to 10 MΩ without upsetting the 0.7RC rule.

Delayed pulses

The circuit symbol for a monostable is shown in Fig. 3.12. The triangle and circle at the input T indicate that it is **triggered** into action by a falling edge. In other words, once T has gone low, it loses all control over the output Q – the pulse will be produced regardless of what happens to the signal at T. The circle at the output Q tell you that the monostable produces a low-going pulse each time that it is triggered. The duration of the pulse is written in the box.

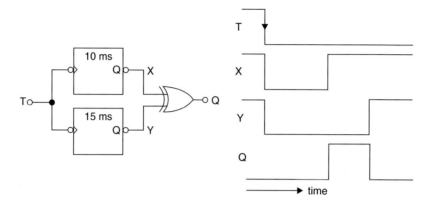

Fig 3.12 A pair of monostables connected to produce a delayed pulse.

The EOR gate in Fig. 3.12 can only go high when its inputs are different. The outputs of both monostables go low at the same time when triggered by a falling edge at T, so Q stays low. However, 10 ms after the arrival of the edge at T the signal at X returns high, making it different from Y, allowing Q to go high. After a further 5 ms, Y also returns high and Q goes low again. The overall result is a single 5 ms high-going pulse which starts 10 ms after a falling edge arrives at T.

Continuous pulses

A pair of monostables can be used to make a system which triggers itself into producing a continuous train of pulses. The arrangement is shown on Fig. 3.13.

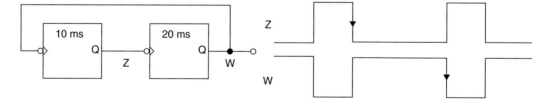

Fig 3.13 A pair of monostables can produce pulses without an external signal.

The monostables need to produce **high-going** pulses, so that when each pulse ends it triggers the other monostable. (The addition of a NOT gate to the output of a low-going monostable is all that is required.) Once the system has started to produce pulses, it will carry on doing so until the power supply has been disconnected. There is no guarantee, however, that the system will start to oscillate spontaneously when it is powered up again! There are other ways of constructing oscillators which will always start up, but the system of Fig. 3.13 has the advantage that the duration of each of the high-going and low-going pulses is set by its own RC network, allowing them to be altered independently.

Digital pulses

3.2 Oscillators

Take another look at the system of Fig. 3.13. It has no input, only an output. The output of each block feeds into the input of the other block in such a way as to keep changing the signals at the outputs. This feedback technique is employed in all **oscillators**: systems which produce signals that vary with time in a regular way. Oscillators are indispensible parts of electronic systems which need to keep track of time (such as clocks) or coordinate the action of different component circuits of a system (such as a computer). The signal produced by the pair of monostables is a **square wave** which continually swaps between 5 V to 0 V. In each cycle of oscillation, the output W is high for 20 ms and low for 10 ms, giving it a **period** of 30 ms and a **mark–space ratio** of 2:1.

Relaxation oscillator

There are many different ways of making circuits oscillate. The simplest, and possibly most reliable, oscillator circuit is shown in Fig. 3.14. It is called a **relaxation oscillator**.

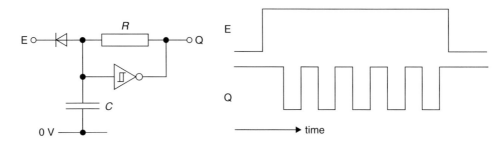

Fig 3.14 A relaxation oscillator based on a Schmitt trigger NOT gate.

As you can see from the timing diagram, the output can only oscillate when the E input is high. (E is short for **enable**.) The oscillating output is a square wave with a mark–space ratio of 1:1 whose **period** (the duration of one cycle of the oscillation) is given by this formula.

$$T = 0.5RC$$

T is the **period** of the square wave in seconds (s), the time taken for just one cycle of the oscillation. R is the resistance in ohms (Ω) and C is the capacitance in farads (F) of the RC network.

Hysteresis

The component at the heart of a relaxation oscillator is a Schmitt trigger NOT gate. It has the normal behaviour of a NOT gate, changing the state of the digital signal at its input. Schmitt triggers also have the property of **hysteresis**, a variable threshold voltage which depends on the state of the gate's output. This is shown in the two transfer characteristics of Fig. 3.15.

Look at the top transfer graph. The input rises from 0 V. At the threshold voltage of 2.8 V it is recognized as a high signal and the output drops low. The bottom graph shows what happens when the input drops from 5 V. Only when it gets to 2.2 V is it recognized as a low signal.

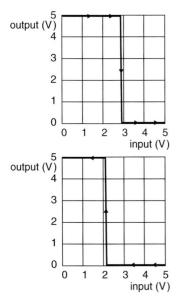

Fig 3.15 The threshold voltage of a Schmitt trigger is different for high and low signals.

OCR Electronics for AS

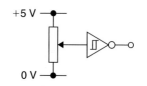

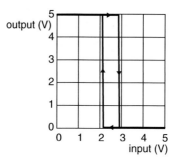

Fig 3.16 Transfer characteristic of a 40106 Schmitt trigger NOT gate.

Fig 3.17 Holding the enable input low forces the output high.

Schmitt trigger

The overall transfer characteristic for a 40106 Schmitt trigger NOT gate is shown in Fig. 3.16. The threshold voltages vary somewhat from one integrated circuit to the other, but are typically at 2.2 V and 2.8 V. These are known as the **trip points** of the gate. Each time the input goes past a trip point, the state of the output changes very rapidly, giving a very clean rising or falling edge.

Why it oscillates

So how does a relaxation oscillator work? Let's start off by considering what happens when the enable input is held low (Fig. 3.17). The diode is forward biased, so the input of the gate is only 0.7 V, making its output 5 V. There is a steady flow of charge from the output, through the resistor and the diode. The voltage across the capacitor is a steady 0.7 V.

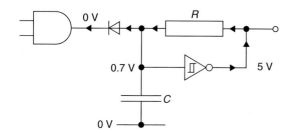

Now suppose that the enable input goes high (Fig. 3.18). The diode is immediately put in reverse bias, so there is no more flow of charge through it. Instead, it flows onto the top plate of the capacitor, raising its voltage as time goes on. The rate of increase of voltage depends on the time constant RC of the RC network – increasing the time constant slows down the voltage rise.

Fig 3.18 A high output charges up the capacitor through the resistor.

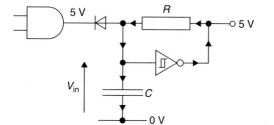
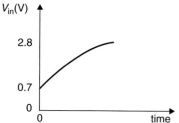

As soon as the input of the NOT gate reaches the **upper trip point** of 2.8 V, the output dives down to 0 V (Fig. 3.19). The flow of charge in the resistor changes direction immediately, discharging the capacitor, causing the voltage at the gate's input to drop as time goes on. Of course, once the voltage at the gate's input reaches the **lower trip point**, the output goes high once more and the capacitor starts to charge up again.

Fig 3.19 A low output discharges the capacitor through the resistor.

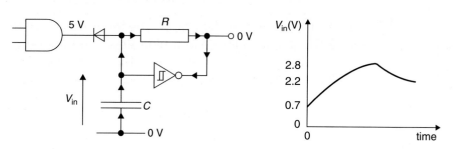

Digital pulses

Period

The graphs of Fig. 3.20 show how the voltages at the input and output of a relaxation oscillator change with time. The period T of the output signal is calculated as follows.

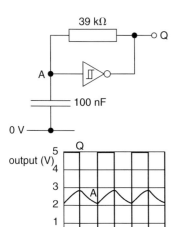

Fig 3.20 A relaxation oscillator in action.

$T = ?$

$R = 39 \text{ k}\Omega$

$C = 100 \text{ nF}$

$T = 0.5RC = 0.5 \times 39 \times 10^3 \times 100 \times 10^{-9}$

$= 2 \times 10^{-3} \text{ s or 2 ms}$

So the output is low for 1 ms and high for 1 ms in each cycle of oscillation. Notice that the resistor value is quite high. This is because the current at the output of a NOT gate is limited to about a milliamp. This means that you can just about get away with a 4.7 kΩ resistor, but it is better to always go for values above 10 kΩ.

Frequency

Oscillators are normally specified by their **frequency**. This is the number of cycles of oscillation that they go through in a second. It can be calculated from the period with this formula.

$$f = \frac{1}{T}$$

The frequency f is measured in hertz (Hz) when the period T is measured in seconds (s). The frequency of the oscillator in Fig. 3.20 can be calculated as follows.

$f = ?$

$T = 2 \text{ ms}$

$f = \frac{1}{T} = \frac{1}{2 \times 10^{-3}} = 5 \times 10^2 \text{ Hz or 0.5 kHz}$

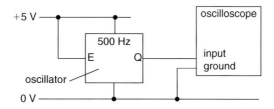

Fig 3.21 Connecting an oscillator to an oscilloscope.

Since each cycle of oscillation only lasts for 2 ms, you are unable to measure its duration with a voltmeter. You need to use an **oscilloscope**. You connect it to the input of the oscillator just like you would connect a voltmeter (Fig. 3.21) and the result is a **trace** on a screen (Fig. 3.22).

The screen trace is essentially a voltage–time graph of the signal at the oscilloscope input. The vertical scale can be set to one of a number of different values, ranging perhaps from 1 mV per division to 50 V per division. In Fig. 3.22, the vertical scale is set to 2 V/div, with 0 V exactly halfway down the screen. The maximum voltage of the signal is 2.5 divisions above the centre of the screen, so its value is $2.5 \times 2 = 5$ V.

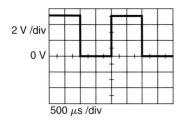

Fig 3.22 Screen trace for the oscillator in Fig. 3.20.

The horizontal scale of the trace is set by the **timebase** of the oscilloscope. For Fig. 3.22, this is set at 500 μs per division. The falling edges of the trace are four divisions apart on the screen, giving a period of $4 \times 500 \times 10^{-6} = 2 \times 10^{-3}$ s or 2 ms.

OCR Electronics for AS

Making sound

Feed the output of an oscillator into a **loudspeaker** and, if the frequency is in the range 16 Hz to 16 kHz, you get a sound. Loudspeakers generally have a low resistance, typically 8 Ω, so you need to place a driver after the oscillator output as shown in Fig. 3.23.

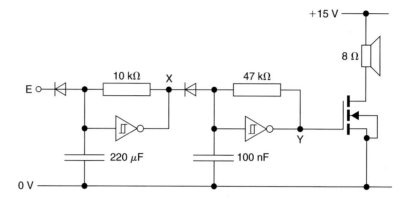

Fig 3.23 Using oscillators to make a sound.

The power delivered to the speaker gives you some idea of the volume of sound it generates. When Y is high, the MOSFET driver will be on, sinking current from the loudspeaker.

$I = ?$

$V = 15$ V

$R = 8$ Ω

$I = \dfrac{V}{R} = \dfrac{15}{8} = 1.9$ A

While Q is high, the power delivered to the loudspeaker is as follows.

$P = ?$

$V = 15$ V

$I = 1.9$ A

$P = IV = 1.9 \times 15 = 28$ W

Of course, when Y is low, the MOSFET is off, delivering no power to the loudspeaker. So the average power of the loudspeaker will be only 14 W – quite a lot of sound.

Oscillated oscillators

You will notice that the circuit in Fig. 3.23 contains two oscillators. The low frequency one on the left is on when E is at +5 V, and it is used to turn the high frequency one on the right on and off. The result is bursts of 430 Hz sound at intervals of half a second – a useful alarm signal. This is summarized in the timing diagram of Fig. 3.24.

Fig 3.24 One oscillator switches the other one on and off.

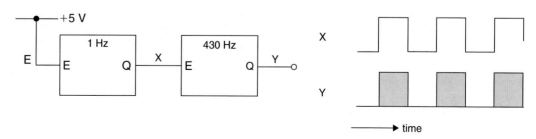

Digital pulses

Questions

3.1 Single spikes

Assume that the logic gates in these questions have threshold voltages of +2.5 V.

1. This question is about a monostable whose output Q falls low for 50 ms each time it is triggered by a falling edge at its input terminal T.

   (a) Draw a timing diagram to represent the behaviour of the monostable.

   (b) Draw a circuit diagram to show how the monostable can be made from NAND gates and an RC network.

   (c) Select component values to produce the required output pulse. Use calculations to justify your choice.

2. This question is about the circuit in Fig. Q3.1. The input signal is a square wave alternating between +5 V and 0 V at a frequency of 0.125 Hz.

   (a) Calculate the period of the input signal.

   (b) Calculate the time constant of the RC network.

   (c) Sketch a voltage–time graph for two cycles of the input signal, marking important values on the axes.

   (d) Sketch a voltage–time graph for the output signal on the same axes as (c).

   (e) Explain how a diode can be used to suppress negative spikes at the output of the circuit.

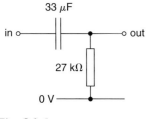

Fig Q3.1

3.2 Oscillators

1. A particular Schmitt trigger NOT gate has the following properties:
   - The output saturates at +0.5 V or +4.5 V.
   - It has trip points at +2.0 V and +3.0 V.

   (a) Draw a transfer characteristic for the NOT gate, to show how the output voltage depends on the input voltage.

   (b) Draw a circuit diagram for a relaxation oscillator based on the NOT gate.

   (c) Sketch voltage–time graphs on the same axes for the two cycles of the signals at the input and output of the NOT gate.

   (d) Explain why the system oscillates.

2. A relaxation oscillator is required to feed a signal at 440 Hz into a loudspeaker via a MOSFET driver.

   (a) Draw a circuit diagram for the system.

   (b) Explain the function of the MOSFET driver.

   (c) Select component values to give the required frequency. Use calculations to justify your choice of values.

   (d) Explain how the system can be adapted so that the oscillator can be turned off and on by a signal from another logic system.

3. This question is about the oscilloscope trace in Fig. Q.3.2.

   (a) The timebase is set to 20 μs per division. Calculate the frequency of the displayed signal.

   (b) The vertical amplifier is set to 0.5 V per division. Calculate the highest and lowest voltages of the signal.

   (c) Sketch the trace on the screen if the timebase is changed to 10 μs per division and the vertical amplifier is changed to 1.0 V per division.

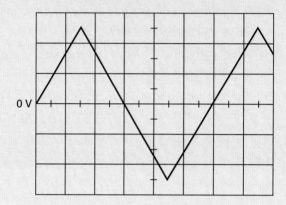

Fig Q3.2

Learning summary

By the end of this chapter you should be able to:

- use an RC network to generate spikes from rising or falling edges
- know the transfer characteristics of a monostable
- construct a monostable from NAND gates and an RC network
- use an RC network and a Schmitt trigger NOT gate to make a relaxation oscillator
- know the transfer characteristics of a relaxation oscillator
- use an oscilloscope to measure the amplitude and frequency of a square wave

CHAPTER 4

Logic systems

4.1 Truth tables

You have already met the six different types of logic gate in Chapter 1. Each gate has its own unique truth table which summarizes its behaviour. For example, the output of an AND gate is only high when both its inputs are high. Simple systems with two inputs and one output can sometimes be designed by selecting a single appropriate logic gate, but these gates are more often used as building blocks for complex systems with many inputs and outputs. In fact, the real power of logic gates is only unleashed when they are connected together as a **logic system**. By feeding the outputs of logic gates into the inputs of other logic gates, you can make systems which have can any truth table you like. For example, consider the two logic systems shown in Fig. 4.1, each made from a different network of logic gates.

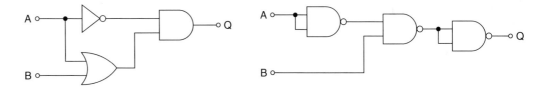

Fig 4.1 Two different logic systems with something in common.

B	A	Q
0	0	0
0	1	0
1	0	1
1	1	0

Both logic systems have the same truth table. Here it is.

So which system should you use? After all, they have the same truth table and are identical as far as their function is concerned. The circuit on the left, however, will need to be assembled from three integrated circuits, one for each type of logic gate. Furthermore, most of the gates in each integrated circuit won't be used, making the circuit economically inefficient. The circuit on the right only uses one type of gate, so only one integrated circuit is needed, thereby taking up less space on the printed circuit board, as well as costing less to build. So the circuit on the right is better.

System behaviour

You can always work out the behaviour of a logic system by working out its truth table. It may not be a very practical method for logic systems which have more than a few inputs, but it is perfectly acceptable for small systems, such as the one shown in Fig. 4.2.

Fig 4.2 A small logic system.

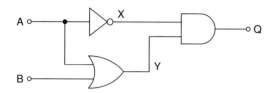

So how do you **analyse** the small system of Fig. 4.2 with a truth table? Start off by writing down a truth table which has all of the different **states** of the inputs. Each state is numbered, starting from 0. The signals at B and A are chosen such that the **binary word** BA has the same value as the state.

state	B	A
0	0	0
1	0	1
2	1	0
3	1	1

Now add columns to show the processing signals at X and Y. X is the output of a NOT gate whose input is A, so X is always the opposite to A. Y is the output of an OR gate, so it is high whenever A or B are high.

state	B	A	X	Y
0	0	0	1	0
1	0	1	0	1
2	1	0	1	1
3	1	1	0	1

Finally add a column for the output Q. The AND gate output can only be high when both X and Y are high. This only happens in state 2, so Q is 0 except for state 2.

state	B	A	X	Y	Q
0	0	0	1	0	0
1	0	1	0	1	0
2	1	0	1	1	1
3	1	1	0	1	0

Logic systems

Boolean algebra

There is another way to analyse a logic system. It uses a notation called **Boolean algebra** and is much more practical than a truth table for large systems. The table below shows the algebraic representation of the three basic logic gates.

logic gate	algebra	description	behaviour
NOT	$Q = \bar{A}$	Q equals NOT A	output is opposite to input
OR	$Q = A + B$	Q equals A OR B	output only low when both inputs are low
AND	$Q = A.B$	Q equals A AND B	output only high when both inputs are high

Let's use this to analyse the behaviour of the small system of Fig. 4.2. Start off by writing down the algebra which represents the behaviour of each gate.

$$X = \bar{A}$$

$$Y = A + B$$

$$Q = X.Y$$

Now combine the three expressions to make one which has the output Q only as a function of the inputs A and B.

$$Q = \bar{A}.(A + B)$$

Although this piece of algebra represents the behaviour of the system, it needs **simplifying** before it makes much sense. For that, you need a few rules.

Brackets

The brackets in a Boolean algebra expression tell you what order the gates are in. So in $Q = \bar{A}.(A + B)$, the brackets tell you that the OR gate has to combine A and B before feeding the result into the AND gate. This means that $Q = (\bar{A}.A) + B$ would be quite a different circuit, with a completely different truth table – it says that an AND gate combines A and $\bar{A}$ before feeding the result into an OR gate.

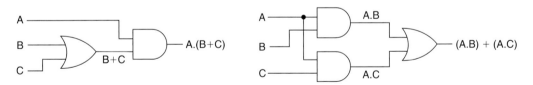

Fig 4.3 These systems have the same truth table.

The two logic systems of Fig. 4.3 have the same truth table, proving the truth of the following useful rule.

$$A.(B + C) = (A.B) + (A.C)$$

Let's apply this rule to the algebra which describes the behaviour of the system of Fig. 4.2. The brackets rule means that $Q = \bar{A}.(A + B)$ can also be represented by $Q = \bar{A}.A + \bar{A}.B$. This is known as a **sum** of two **terms**, a format which, as you will find out, can be easily used to design logic systems as well as to analyse them. Note that the absence of any brackets at all tells you that the AND function in each term must be performed before the terms are combined by the OR function.

AND rules

Here are some useful rules involving the AND function.

$$A.A = A$$
$$A.0 = 0$$
$$A.1 = A$$
$$A.\overline{A} = 0$$

B	A	Q
0	0	0
0	1	0
1	0	0
1	1	1

The first three are obvious from the inspection of the AND gate truth table. If B and A are the same, then only the first and last lines of the table apply, with Q always the same as A. Similarly, if B is always 0, then Q must be 0. Finally, if B is always 1, then Q is the same as A.

The last rule can be understood by using a truth table to analyse the system of Fig. 4.4. The inputs of the gate always have different signals, so the output is always 0.

Fig 4.4 Logic system to show that $A.\overline{A} = 0$.

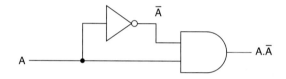

A	$\overline{A}$	$A.\overline{A}$
0	1	0
1	0	0

OR rules

Here are the equivalent rules for the OR function.

$$A + A = A$$
$$A + 0 = A$$
$$A + 1 = 1$$
$$A + \overline{A} = 1$$

B	A	Q
0	0	0
0	1	1
1	0	1
1	1	1

The first three rules are pretty obvious if you study the OR gate truth table.

The system of Fig. 4.5 shows the truth of the last rule. One of the inputs of the gate will always be high, so its output must always be 1.

Fig 4.5 Logic system to show that $A + \overline{A} = 1$.

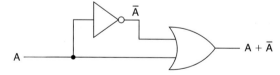

A	$\overline{A}$	$A + \overline{A}$
0	1	1
1	0	1

Logic systems

Simplifying

The rules about brackets, the OR function and the AND function can be used to **simplify** the expressions representing the behaviour of logic systems.

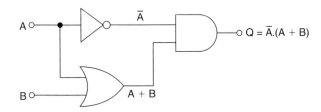

Fig 4.6 This system obeys the expression $Q = \bar{A}.B$.

Let's use the familiar logic system of Fig. 4.6 as an example. The arrangement of logic gates tells you that $Q = \bar{A}.(A + B)$. Then simplify it as follows:

1 Remove the brackets: $Q = \bar{A}.A + \bar{A}.B$

2 Use $A.\bar{A} = 0$: $Q = 0 + \bar{A}.B$

3 Use $A + 0 = A$: $Q = \bar{A}.B$

So what does this tell you? It says that Q is a 1 for only one state, when A is 0 and B is 1, as shown in this truth table.

B	A	$Q = \bar{A}.B$
0	0	0
0	1	0
1	0	1
1	1	0

Table to algebra

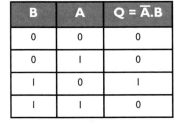

Fig. 4.7 shows the circuit symbol for a useful logic system called a **multiplexer**. It has three inputs (S_0, S_1 and C) and one output (Q).

Fig 4.7 A multiplexer allows only one of a pair of signals through to the output.

Two signals S_0 and S_1 enter from the left. The third control signal C decides which of S_0 or S_1 arrives at the output Q. When $C = 1$, $Q = S_1$. When $C = 0$, $Q = S_0$. This gives the truth table below with eight different input states.

This is how to reduce the table to a single Boolean algebra expression.

1 Identify the states where $Q = 1$.

2 Use the AND and NOT functions to create **terms** which represent each of these states.

3 Use the OR function to combine the terms into the final expression.

Let's apply these rules to the truth table opposite. Q is high for state 2, when $C = 0$, $S_0 = 1$ and $S_1 = 0$, therefore one term is $\bar{C}.S_0.\bar{S_1}$. Q is also high for states 3, 5 and 7, giving three more terms: $\bar{C}.S_0.S_1$, $C.\bar{S_0}.S_1$ and $C.S_0.S_1$. The overall expression is the sum of all four terms:

$$Q = \bar{C}.S_0.\bar{S_1} + \bar{C}.S_0.S_1 + C.\bar{S_0}.S_1 + C.S_0.S_1$$

state	C	S_0	S_1	Q
0	0	0	0	0
1	0	0	1	0
2	0	1	0	1
3	0	1	1	1
4	1	0	0	0
5	1	0	1	1
6	1	1	0	0
7	1	1	1	1

OCR Electronics for AS

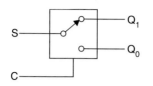

Fig 4.8 A demultiplexer directs an incoming signal to one or other output.

C	S	Q_1	Q_0
0	0	0	0
0	1	0	1
1	0	0	0
1	1	1	0

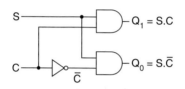

Fig 4.9 A logic system for a demultiplexer.

4.2 Logic system design

So far, Boolean algebra has been introduced as a type of shorthand. It allows you to replace a large truth table with a single line of symbols. It really comes into its own when you need to design a logic system from a truth table.

Demultiplexer

A demultiplexer has two inputs (S and C) and two outputs (Q_0 and Q_1), as shown in Fig 4.8.

The signal at C decides where the signal at S ends up. When C = 1, Q_1 = S and Q_0 = 0. When C = 0, Q_0 = S and Q_1 = 0.

Here are the steps to converting this specification into a working circuit of basic logic gates:

1 Write out a truth table showing output signals for all possible input states.

2 State Boolean expressions for each output in terms of only the inputs.

$$Q_1 = C.S$$
$$Q_0 = \overline{C}.S$$

3 Draw the logic system as shown in Fig. 4.9.

Notice how the expressions for the outputs come from the truth table. Q_0 is 1 only when C = 0 and S = 1. This is the same as saying that $\overline{C}$ = 1 and S = 1. Hence the expression $Q_0 = \overline{C}.S$, which only needs a NOT gate and an AND gate for its implementation in hardware.

Multiplexer

As you found out on the previous page, a multiplexer can be represented with this expression.

$$Q = \overline{C}.S_0.\overline{S_1} + \overline{C}.S_0.S_1 + C.\overline{S_0}.S_1 + C.S_0.S_1$$

Fig. 4.10 on the next page shows how this can be implemented by basic logic gates. Notice that:

- Q is generated by a four-input OR gate
- each of the four terms is generated by a three-input AND gate
- only three NOT gates are required
- the output of each logic gate has its associated Boolean expression.

Logic systems

Fig 4.10 An inelegant logic system for a multiplexer.

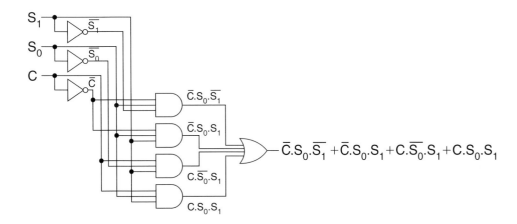

Fig. 4.11 shows how the multiple input gates shown in Fig. 4.10 can be implemented from basic dual-input gates.

Fig 4.11 Making large gates from small ones.

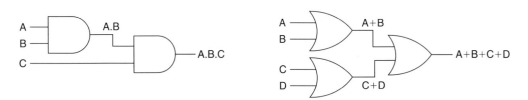

Simplifying expressions

Although the multiplexer circuit shown above works, it is not a very clever way of implementing the specification. It uses far too many basic logic gates: eight AND gates, three OR gates and three NOT gates. That's four integrated circuits. It would have been better to **simplify** the Boolean expression first as follows:

1 Carefully inspect the four terms in the expression obtained from the truth table.

$$Q = \overline{C}.S_0.\overline{S_1} + \overline{C}.S_0.S_1 + C.\overline{S_0}.S_1 + C.S_0.S_1$$

2 Insert brackets where terms have parts in common.

$$Q = \overline{C}.S_0.(\overline{S_1} + S_1) + C.S_1.(\overline{S_0} + S_0)$$

3 Use the rules $\overline{A} + A = 1$ and $A.1 = A$ to eliminate the brackets.

$$Q = \overline{C}.S_0 + C.S_1$$

Fig. 4.12 shows that this expression only needs four logic gates for its implementation.

Fig 4.12 A better logic system for a multiplexer.

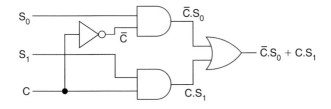

NOR function

B	A	Q
0	0	1
0	1	0
1	0	0
1	1	0

Here is the truth table for a NOR gate.

Q is only high when both B and A are low, so the output of the gate is represented by the expression $Q = \overline{B}.\overline{A}$. The circuit on the left of Fig. 4.13 shows how this can be implemented in basic logic gates.

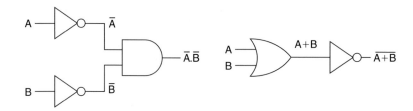

Fig 4.13 Two logic systems which behave like NOR gates.

It is interesting to note that the circuit on the right of Fig. 4.13 also has the truth table of a NOR gate.

B	A	B + A	$\overline{B + A}$
0	0	0	1
0	1	1	0
1	0	1	0
1	1	1	0

This leads to **De Morgan's Theorem**.

$$\overline{A + B} = \overline{A}.\overline{B}$$

This is useful for reducing Boolean expressions into sums of terms, the best form for implementation in basic logic gates.

Race Hazard and Redundancy

The **Race Hazard Theorem** allows you to add an extra term to an expression.

$$X.Y + \overline{X}.Z = X.Y + \overline{X}.Z + Y.Z$$

Notice how one term on the left contains X and the other contains $\overline{X}$, and the extra term on the right is made by ANDing the other parts of the terms on the left.

The **Redundancy Theorem** does the opposite. It allows you to remove a long term which has a shorter term buried inside it.

$$X + X.Y = X$$

As you will see, these theorems can be used in pairs to simplify Boolean expressions.

Logic systems

NAND function

Here is the truth table for a NAND gate.

B	A	Q
0	0	1
0	1	1
1	0	1
1	1	0

Q is high for three states, so its expression contains three terms.

$$Q = \bar{B}.\bar{A} + \bar{B}.A + B.\bar{A}$$

This can be simplified as follows:

1 Use brackets to extract common parts of the first two terms.

$$Q = \bar{B}.(\bar{A} + A) + B.\bar{A}$$

2 Use the rules $\bar{A} + A = 1$ and $A.1 = A$ to eliminate the brackets.

$$Q = \bar{B} + B.\bar{A}$$

3 Use the Race Hazard Theorem ($X.Y + \bar{X}.Z = X.Y + \bar{X}.Z + Y.Z$) to add an extra term.

$$Q = \bar{B} + B.\bar{A} + \bar{A}$$

4 Finally, use the Redundancy Theorem ($X + X.Y = X$) to remove the middle term.

$$Q = \bar{B} + \bar{A}$$

The circuit on the left of Fig. 4.14 shows how this can be implemented in basic logic gates.

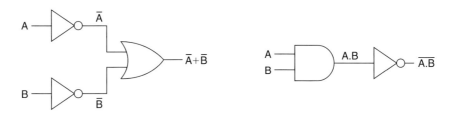

Fig 4.14 Two logic systems which behave like NAND gates.

It is interesting to note that the circuit on the right of Fig. 4.14 also has the truth table of a NAND gate.

B	A	B.A	$\overline{B.A}$
0	0	0	1
0	1	0	0
1	0	0	0
1	1	1	0

This leads to another version of **De Morgan's Theorem**.

$$\overline{B.A} = \bar{B} + \bar{A}$$

Both versions of this theorem require you to do the same thing:

1 Break the bar over a pair of variables.

2 Change the operator linking the variables (swap AND for OR and vice versa).

OCR Electronics for AS

EOR function

The circuit in Fig. 4.15 shows one particularly useful way of assembling an EOR gate, using all four NAND gates in an integrated circuit.

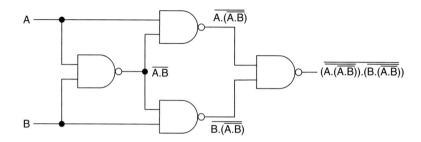

Fig 4.15 A NAND gate circuit with the truth table of an EOR gate.

You have to analyse the behaviour of a logic circuit by starting at the left and building up expressions for each gate as you move to the right, until you end up at the output. In this case, the result is this unwieldy expression:

$$Q = \overline{(A.\overline{(A.B)}).(B.\overline{(A.B)})}$$

You need to apply De Morgan's Theorem to reduce this to a sum of terms.

$$Q = \overline{A.\overline{(A.B)}} + \overline{B.\overline{(A.B)}}$$

Now use the fact that $\overline{\overline{X}} = X$ to get rid of the double inversion of each term.

$$Q = A.\overline{(A.B)} + B.\overline{(A.B)}$$

Now apply De Morgan's Theorem for a second time.

$$Q = A.(\overline{A} + \overline{B}) + B.(\overline{A} + \overline{B})$$

Eliminate the brackets to give this sum of four terms.

$$Q = A.\overline{A} + A.\overline{B} + B.\overline{A} + B.\overline{B}$$

Finally, use $X.\overline{X} = 0$ and $X + 0 = X$ to get rid of the first and last terms.

$$Q = A.\overline{B} + \overline{A}.B$$

This matches the truth table for an EOR gate, shown below.

B	A	Q
0	0	0
0	1	1
1	0	1
1	1	0

Logic systems

4.3 Only NAND gates

A long time ago, the electronics industry decided that it would be a good idea to make large logic systems out of just one type of gate. This has several advantages:

- It often minimizes the number of integrated circuits you need to use.
- Logic systems take up less space on the printed circuit board.
- Very few integrated circuits have unused gates.
- More of that type of gate are produced, lowering their manufacturing cost.
- You do not have to stock and store many different integrated circuits, reducing costs.

For example, Fig. 4.16 shows how an EOR gate can be implemented with basic logic gates. It needs three integrated circuits (AND, OR and NOT), only using five out of the fourteen gates in those packages. Contrast that with the NAND gate circuit shown in Fig. 4.15, where all four gates in a single integrated circuit are used.

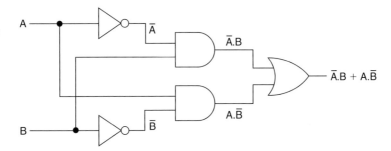

Fig 4.16 EOR gate implemented from basic logic gates.

It is possible to build any logic system out of just NAND gates because they can be used to make all three basic logic gates (NOT, AND and OR). So it makes sense, for all of the reasons listed above, to design many logic systems out of just NAND gates. In fact, NOR gates can also be used to make all three basic logic gates, but the economic advantages only happen if everyone chooses to use the same logic gate – and the NAND gate got lucky at the expense of NOR gates!

NOT gates

There are two ways of making a NOT gate out of a NAND gate. Either hold one of the inputs high all of the time or join both inputs together. Both strategies are shown in Fig. 4.17.

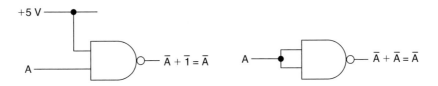

Fig 4.17 Two ways of making a NOT gate from a NAND gate.

Here is one of the Boolean expressions for a NAND gate:

$$Q = \overline{A} + \overline{B}$$

So the circuit on the left of Fig. 4.17 obeys $Q = \overline{A} + \overline{1} = \overline{A} + 0 = \overline{A}$. The circuit on the right obeys $Q = \overline{A} + \overline{A} = \overline{A}$, so it also acts like a NOT gate.

OR gates

B	A	Q
0	0	0
0	1	1
1	0	1
1	1	1

Here is the truth table for an OR gate.

Q is high when A is high or when B is high, so it obeys the expression Q = A + B. You can implement this two-term expression with NAND gates as follows:

1 Use a two-input NAND gate to generate Q.

2 Use Q = $\overline{X}$ + $\overline{Y}$ to label the inputs with the inverse of each term, i.e. $\overline{A}$, $\overline{B}$.

3 Use a pair of NAND gates to generate $\overline{A}$ and $\overline{B}$.

The result of these operations is shown in Fig. 4.18.

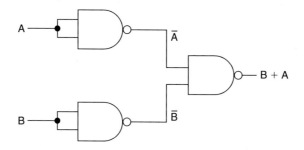

Fig 4.18 An OR gate made from NAND gates.

AND gates

B	A	Q
0	0	0
0	1	0
1	0	0
1	1	1

An AND gate obeys the single term expression Q = A.B. This should be obvious from its truth table opposite.

You implement this with a pair of NAND gates as follows:

1 Use a NAND gate with both inputs joined together to generate Q.

2 Label the input with the inverse of the single term. i.e. $\overline{A.B}$.

3 Use a single two-input NAND gate to generate $\overline{A.B}$.

The result of these operations is shown in Fig. 4.19.

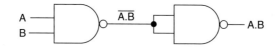

Fig 4.19 An AND gate made from NAND gates.

Logic systems

Large systems

This section takes you through an example of using only NAND gates to design a logic system. The system has three inputs (C, B and A) and three outputs (X, Y and Z), as shown in Fig. 4.20.

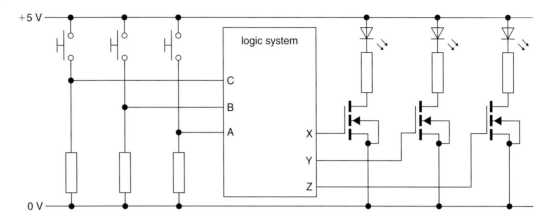

Fig 4.20 The number of glowing LEDs is the same as the number of closed switches.

The specification is as follows: the number of glowing LEDs must be the same as the number of switches being pressed.

As always, you should start off with a truth table, bearing in mind that closing a switch pulls an input high and a glowing LED requires a high output.

C	B	A	X	Y	Z
0	0	0	0	0	0
0	0	1	1	0	0
0	1	0	1	0	0
0	1	1	1	1	0
1	0	0	1	0	0
1	0	1	1	1	0
1	1	0	1	1	0
1	1	1	1	1	1

Then you write down expressions for each output in terms of the inputs. Let's start with X. It is high for seven states, so its expression should have seven terms. This is going to be awkward. So it is a better idea to write an expression which tells you the input conditions required to make X go low instead. This has only one term.

$$\overline{X} = \overline{C}.\overline{B}.\overline{A}$$

If you invert both sides and apply De Morgan's Theorem, you obtain an expression for X.

$$\overline{\overline{X}} = X = \overline{\overline{C}.\overline{B}.\overline{A}} = C + B + A$$

This can be implemented straight away with a single three-input NAND gate and three two-input ones. (NAND gates come with two, three or four inputs.) (Fig 4.21)

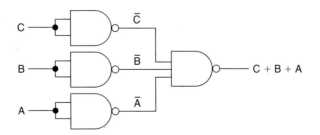

Fig 4.21 The logic subsystem for X.

The expressions for Y and Z are more straightforward.

$$Y = \overline{C}.B.A + C.\overline{B}.A + C.B.\overline{A} + C.B.A$$

$$Z = C.B.A$$

OCR Electronics for AS

Y subsystem

The expression for Y can be simplified as follows:

1 Use brackets to link the first and last terms.

$$Y = \overline{C}.B.A + C.\overline{B}.A + C.B.\overline{A} + C.B.A$$

$$Y = B.A.(\overline{C} + C) + C.\overline{B}.A + C.B.\overline{A}$$

2 Use the rules $\overline{A} + A = 1$ and $A.1 = A$ to eliminate the brackets.

$$Y = B.A + C.\overline{B}.A + C.B.\overline{A}$$

3 Apply the Race Hazard Theorem to the first two terms.

$$Y = B.A + C.\overline{B}.A + A.C + C.B.\overline{A}$$

4 Use the Redundancy Theorem to eliminate the second term.

$$Y = B.A + A.C + C.B.\overline{A}$$

5 Apply the Race Hazard Theorem to the last two terms.

$$Y = B.A + A.C + C.B.\overline{A} + C.B$$

6 Use the Redundancy Theorem to eliminate the third term.

$$Y = B.A + A.C + C.B$$

Fig. 4.22 shows how this can be implemented in NAND gates. Notice that the number of terms in the expression determines the number of inputs to the gate which generates Y.

Fig 4.22 The logic subsystem for Y.

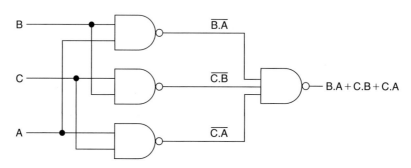

Z subsystem

The expression for Z has only one term.

$$Z = C.B.A$$

This means that the NAND gate which generates it only needs one input. A three-input NAND gate can then be used to generate the signal $\overline{C.B.A}$ at the single input as shown in Fig. 4.23.

Fig 4.23 The logic subsystem for Z.

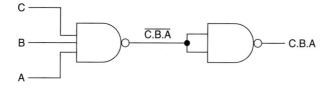

Logic systems

Analysing logic systems

Take a look at the NAND gate logic system of Fig. 4.24.

Fig 4.24 A logic system.

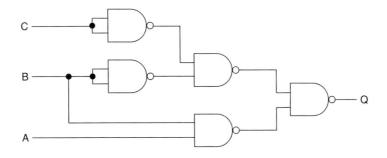

How do you set about using Boolean algebra to work out what it does? You have to start from the inputs and work your way through the gates until you reach the output – exactly the opposite of the way you design a system from an expression. Fig. 4.25 shows what you should have when you have worked your way through the first two layers of gates. Notice that you need to use the idea that NAND gates obey $Q = \overline{X.Y}$ for this stage.

Fig 4.25 The outputs of the first two layers of the system.

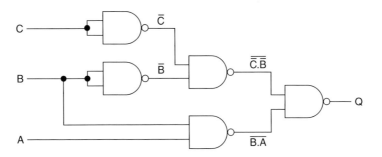

Now use the idea that $Q = \overline{X} + \overline{Y}$ to write down the output of the final gate from the signals at its inputs.

$$Q = \overline{\overline{C.B}} + \overline{\overline{B.A}}$$

Finally, remove the double inversions.

$$Q = \overline{C.B} + B.A$$

This sum of terms can now be used to fill in a truth table. Q must only be high whenever B and A are high OR when both C and B are low.

C	B	A	Q
0	0	0	1
0	0	1	1
0	1	0	0
0	1	1	1
1	0	0	0
1	0	1	0
1	1	0	0
1	1	1	1

OCR Electronics for AS

Questions

4.1 Truth tables

1. This question is about the logic system in Fig. Q4.1.

   (a) Write out a truth table to show how the signals at X, Y, Z and Q depend on the signals at C, B and A.

   (b) Write down Boolean algebra expressions for X, Y and Z in terms of C, B and A.

   (c) Show that the logic system can be replaced with a single two-input logic gate.

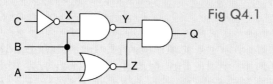

Fig Q4.1

2. The logic system in Fig. Q4.2 has three inputs and two outputs.

   (a) Write down a Boolean algebra expression for X in terms of C, B and A.

   (b) Show that $X = A + C$.

   (c) Write down a Boolean algebra expression for Y in terms of C, B and A.

   (d) Show that $Y = A.(\overline{B} + \overline{C})$.

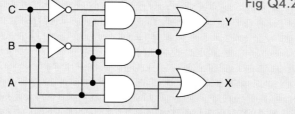

Fig Q4.2

3. A logic system has three inputs, C, B and A. Its output Q obeys this truth table.

   (a) Write down a Boolean algebra expression for Q in terms of C, B and A.

   (b) Show that $Q = A.(\overline{C} + B)$.

   (c) Draw a circuit to implement the logic system, using NOT, AND and OR gates.

4. This question is about the use of brackets in Boolean algebra.

   (a) Draw a logic system which obeys the expression $Q = A.(\overline{B} + \overline{C})$.

   (b) Write out a truth table for the logic system, with a column for the output of each logic gate.

   (c) Write out a truth table for $P = A.\overline{B} + \overline{C}$.

   (d) Eliminate the brackets in the expression for Q and draw a logic system for the new expression.

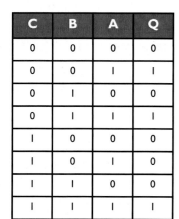

4.3 Only NAND gates

1. A NAND gate has two inputs A, B and one output Q.

   (a) Write the truth table for a NAND gate.

Logic systems

   (b) Write down **two** different Boolean expressions which represent the behaviour of a NAND gate.

   (c) Draw circuit diagrams to show how NAND gates can be used to make systems which behave like

   (i) a NOT gate

   (ii) an AND gate

   (iii) an OR gate

   Use Boolean algebra or truth tables to justify your arrangement of NAND gates.

   (d) State the advantages of building logic systems from just NAND gates.

2 A logic system has three inputs, A, B and C, and one output Q. It is represented by the expression $Q = \overline{A}.B.\overline{C} + A.B.\overline{C} + \overline{(B + \overline{C})}$.

   (a) Use the theorems of Boolean algebra to show that $Q = B.\overline{C} + \overline{B}.C$.

   (b) Show how the logic system can be implemented with NOT, AND and OR gates.

   (c) Show how the logic system can be implemented with just NAND gates.

3 The logic system in Fig. Q4.3 is a one-of-four decoder. It obeys this truth table.

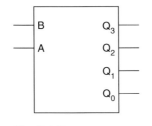

Fig Q4.3

B	A	Q_3	Q_2	Q_1	Q_0
0	0	1	1	1	0
0	1	1	1	0	1
1	0	1	0	1	1
1	1	0	1	1	1

   (a) Write down Boolean expressions for each of the outputs.

   (b) Show that $Q_2 = \overline{B.\overline{A}}$.

   (c) Draw a circuit to show how the decoder can be implemented from just six NAND gates.

   (d) By drawing a circuit to show how the decoder can be implemented from NOT, AND and OR gates, explain the advantages of only using NAND gates to implement logic systems.

Learning summary

By the end of this chapter you should be able to:

- use a truth table to analyse the behaviour of a logic system
- use Boolean algebra to represent logic systems and truth tables
- use the rules of Boolean algebra to design logic systems
- make AND, OR and NOT gates from just NAND gates
- design logic systems from just NAND gates

CHAPTER 5

Storing signals

5.1 Bistables

All of the systems that you met in the last chapter have one thing in common. The state of their outputs depends on the current state of their inputs. Change an input signal and within a fraction of a microsecond the output signal changes accordingly. The systems in this chapter are quite different. Their output signals are determined by not only the current state of their inputs, but also what those inputs were in the past. They have the property of **memory**, the important ability to store events that happened previously. Computers would be impossible without them.

Two stable states

Take a pair of NOT gates and connect them to each other as shown in Fig. 5.1. What will Q be when you switch on the power supply?

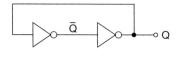

Fig 5.1 A NOT gate bistable.

You cannot use Boolean algebra to find the answer because the circuit has no input. So all you can do is guess what Q will be and then check to see if your guess is consistent.

Suppose you guess that Q is high. Then $\overline{Q}$ must be low, keeping Q high. So you guessed right first time; Q must be high. Hang on, though. Suppose that Q is low. Doesn't that make $\overline{Q}$ high? Isn't that enough to keep Q low?

So Q can be high or low. The circuit has two **stable** states, hence its title of **NOT gate bistable**. Whatever state it is in, it will continue to stay there. But you have no way of knowing which state it will be in when the power is switched on. Without an input terminal, there is no way that you can influence the state of Q – once it is in a state, the **feedback loop** through the two NOT gates keeps it there until the power is switched off again.

Set and reset

The two output states of all bistable circuits are given special names. When Q is **set**, it is high. When Q is **reset**, it is low. A useful bistable must have inputs which can be used to set and reset it. Fig. 5.2 shows how a couple of push switches can provide such inputs for a NOT gate bistable.

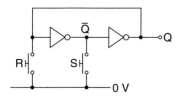

Fig 5.2 This bistable can be set and reset by pressing the switches.

When both switches are open, the circuit is exactly the same as that of Fig. 5.1, so you have no way of working out if it is set or reset. However, pressing the switch labelled R forces the output Q to 0 V (fortunately logic gates are designed to survive this sort of abuse), thereby resetting the bistable. This is shown on the left of Fig. 5.3.

Fig 5.3 Resetting the bistable.

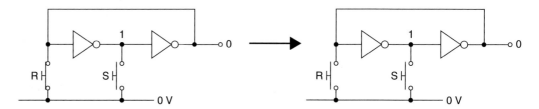

Releasing the switch leaves the bistable in its new state, with the feedback loop keeping Q low. If you do not press any switches, the bistable will remain reset until the power supply is turned off. However, as soon as you press the switch labelled S (as shown on the left in Fig. 5.4), $\overline{Q}$ is forced low by direct connection to the 0 V supply rail. If $\overline{Q}$ is low, then Q must be high, so this action sets the bistable. Releasing the switch leaves the system in this state.

Fig 5.4 Setting the bistable.

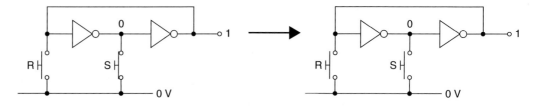

Timing diagrams

Since the behaviour of bistables is time-dependent, it makes sense to illustrate their behaviour as timing diagrams instead of truth tables. Fig. 5.5 shows the sequence of states for the bistable in Fig. 5.2 as it is switched on, reset, set and then reset again.

Fig 5.5 Timing diagram for a NOT gate bistable.

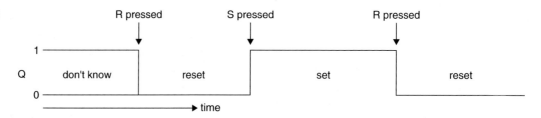

Notice that at the start, before any switches have been pressed, you do not know what Q is. It could be high. It could be low. So it is represented as being both high and low at the same time. Pressing the R switch briefly removes this ambiguity straight away. Q plunges low and stays there. Similarly, as soon as the S switch is pressed, Q goes high immediately and stays there when the switch is released again.

Storing signals

NOR gate bistable

A NOT gate bistable has to be set and reset with switches. A system which can only remember which of two switches was last pressed has limited uses. Switches are mechanical inputs – slow and liable to wear out. Electrical inputs would be much better – faster, nothing to wear out and no need for the presence of a human. The possibilities are endless for a system which can remember which of two electrical inputs was last pulsed high. Such a system, known as a **NOR gate bistable**, is shown with its timing diagram in Fig. 5.6.

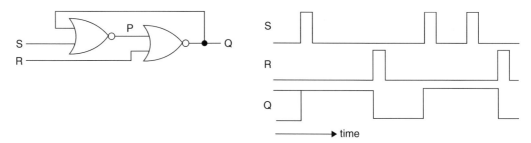

Fig 5.6 Timing diagram for a NOR gate bistable.

The timing diagram summarizes the behaviour of the system. It shows how the state of the output Q depends on the changing signals at the inputs S and R. Initially, both S and R are low, so the state of Q is uncertain. However, as soon as S is pulsed high, Q goes high, remaining high until R is pulsed high. In other words, the state of Q tells you which of the two inputs was last pulsed high.

Active-high

The NOR gate bistable has **active-high inputs**. This means that the inputs have to be **pulsed high** for them to affect the state of the bistable. When both inputs are low, the output remains stable at whatever value (1 or 0) it was previously. You can see this from the pair of circuits shown in Fig. 5.7 and the NOR gate truth table.

B	A	Q
0	0	1
0	1	0
1	0	0
1	1	0

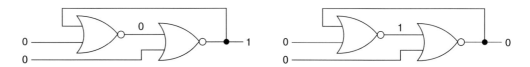

Fig 5.7 The output can be 1 or 0 when the inputs are both 0.

This table summarizes the behaviour of the NOR gate bistable.

S	R	Q
0	0	1 or 0
0	1	0
1	0	1

OCR Electronics for AS

Forbidden states

Notice that the input state SR = 11 is not included in the table on the previous page. It is **forbidden**. This is because you cannot, in practice, go straight from SR = 11 to SR = 00. One of the inputs will always go low before the other, perhaps only for a fraction of a microsecond. If SR goes from 11 through 10 to 00, then Q will be 1 and the bistable will be set. On the other hand, if SR goes from 11 through 01 to 00, then Q will be 0 and the bistable will be reset. So the state SR = 11 is to be avoided since the result of leaving it is liable to be unpredictable.

Algebraic analysis

Since the NOR gate bistable of Fig. 5.6 (repeated below) has two digital inputs as well as an output, the behaviour of the bistable can be analysed with Boolean algebra:

1. Start off by writing down separate expressions for Q and P.

$$Q = \overline{P + R}$$
$$P = \overline{S + Q}$$

2. Apply De Morgan's Theorem to the expression for Q and insert the expression for P.

$$Q = \overline{P}.\overline{R} = \overline{\overline{(S + Q)}}.\overline{R} = (S + Q).\overline{R}$$

3. Finally, remove the brackets to obtain an expression for Q in terms of itself, S and R.

$$Q = S.\overline{R} + Q.\overline{R}$$

If this expression is inserted into a truth table, its meaning becomes clear.

S	R	$Q = S.\overline{R} + Q.\overline{R}$
0	0	$0.\overline{0} + Q.\overline{0} = 0.1 + Q.1 = Q$
0	1	$0.\overline{1} + Q.\overline{1} = 0.0 + Q.0 = 0$
1	0	$1.\overline{0} + Q.\overline{0} = 1.1 + Q.1 = 1$

Fig 5.6 (Repeated from page 69) Timing diagram for a NOR gate bistable.

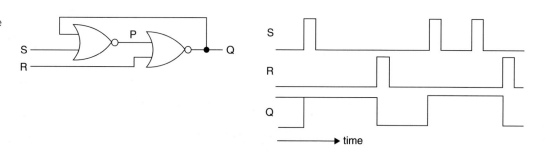

Storing signals

NAND gate bistable

A bistable with **active-low inputs** is shown in Fig. 5.8. As you can see from the timing diagram, the system is stable when both inputs are high (SR = 11), so the inputs have to be pulsed low to affect the state of the output.

Fig 5.8 Timing diagram for a NAND gate bistable.

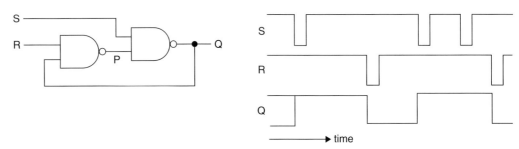

S	R	Q
1	1	1 or 0
0	1	1
1	0	0

The entries for Q in the truth table above can be checked with some Boolean algebra:

1 Write down expressions for the output of each gate in terms of its inputs.

$$Q = \bar{S} + \bar{P}$$

$$P = \overline{R.Q}$$

2 Combine the expressions to eliminate P.

$$Q = \bar{S} + \overline{\overline{R.Q}} = \bar{S} + R.Q$$

3 Consider each input state separately.

S	R	$Q = \bar{S} + R.Q$
1	1	$\bar{1} + 1.Q = 0 + Q = Q$
0	1	$\bar{0} + 1.Q = 1 + Q = 1$
1	0	$\bar{1} + 0.Q = 0 + 0 = 0$

As you can see from the sequence of Fig. 5.9, a NAND gate bistable remembers which of its inputs was last pulled low.

Fig 5.9 Cycling through the stable states of a NAND gate bistable.

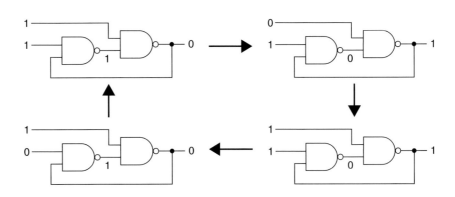

5.2 Latches

The state of a bistable is determined by two inputs. One sets the bistable. The other resets it. As it stands, this is not terribly useful in real systems, such as computers, where you often want to remember the state of an input at a particular time. A system which does this is called a **latch**, and it has the timing diagram shown in Fig. 5.10.

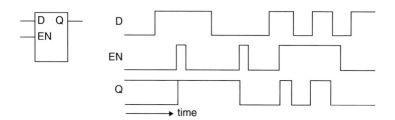

Fig 5.10 Timing diagram for a latch.

As you can see from the timing diagram, the output Q is always the same as the **data input** (D), provided that the **enable input** (EN) is high. In other words, when EN is high, the latch is **transparent** with signals passing straight through it. However, when EN is low, the latch is **frozen** with Q remaining stable regardless of the changes at D.

Data capture

Latches are useful for capturing data. This is how to use a latch to capture one bit of data:

1 Place the bit (1 or 0) to be stored on the data input.

2 Pulse the enable input high to copy the data at D to the output Q.

The stored bit can now be safely read from Q, even when it disappears from D. Of course, this does rely on the data placed at D being stable while EN is high because the bit stored at Q is the state of D at the instant that EN goes back low.

Storing words

A single bit does not hold much information. It can represent just one of two states: open or closed, high or low, in or out, left or right. However, four bits can record a number between 0 and 9 and seven bits can record any letter of the alphabet. So devices which can capture several bits at once and store them are very useful. One such system, the **four-bit latch**, is shown in Fig. 5.11. Each of the four latches lets one bit of the four-bit word placed at the inputs through to the outputs when the enable input is high. As soon as the enable input goes low, the outputs freeze at whatever they where when EN was last high.

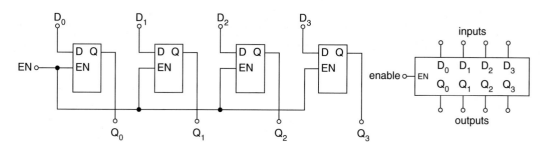

Fig 5.11 A four-bit latch made from one-bit latches.

Storing signals

Inside the latch

Since a latch has to remember its state, it will come as no surprise that it has be built around a bistable. In fact, it is made by using a demultiplexer to generate the signals for a bistable as shown in Fig. 5.12.

Fig 5.12 Inside a latch.

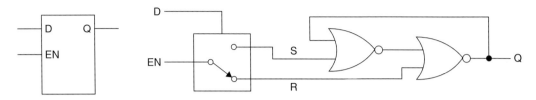

EN	D	S	R
0	0	0	0
0	1	0	0
1	0	0	1
1	1	1	0

So how does it work? The demultiplexer feeds the signal at EN to either S or R, depending on the signal at D. So when EN is high, that signal is passed to either S or R as shown opposite.

The bistable is made from NOR gates, so it has active-high inputs. A high signal at S sets Q high, and a high signal at R resets Q low instead, allowing the system to be transparent; Q follows D. The instant EN goes low, both S and R go low, freezing Q into its latest state. Whatever changes are made to D, they cannot affect S and R because they have to be low when EN is low.

Fine detail

Fig. 5.13 shows how to construct a latch from basic logic gates.

Fig 5.13 A latch made from basic logic gates.

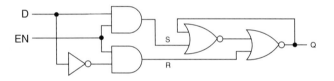

The demultiplexer is shown on the left. Its outputs S and R are represented with these two expressions.

$$S = D.EN$$

$$R = \overline{D}.EN$$

So when EN is 0, both S and R have to be 0 and the bistable is frozen. As soon as EN is 1, then S goes high if D is high, otherwise R goes high. Of course, pushing S or R high results in the bistable being set or reset.

OCR Electronics for AS

Set, reset and enable

It is often useful for a latch to have separate inputs to set and reset the output Q, as well as data and enable inputs to capture data. Such a latch, with its timing diagram, is shown in Fig. 5.14.

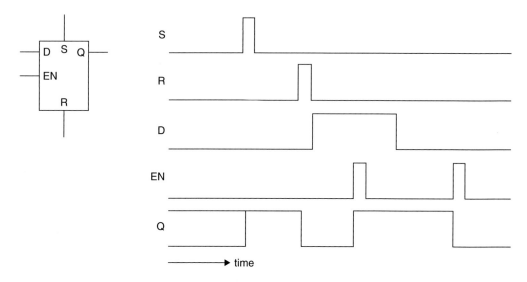

Fig 5.14 A latch with separate set and reset pins.

The S and R inputs are active-high, and their action takes precedence over any signals present at D and EN. In other words, the D and EN inputs can only have any effect on the state of Q when both S and R are low.

NAND gates

The latch of Fig. 5.14 can be assembled from just seven logic gates as shown in Fig. 5.15.

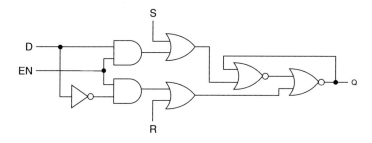

Fig 5.15 The latch in Fig. 5.14 made from a mixture of logic gates.

The OR gates allow the signals at S and R to set and reset the bistable, regardless of the signals at D and EN. The NAND-gate-only version of Fig. 5.16 uses more gates, but takes up less space on the chip when it is built as an integrated circuit.

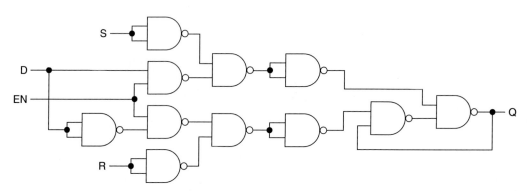

Fig 5.16 Building a latch from just NAND gates.

Storing signals

5.3 Flip-flops

Digital systems which rely on remembering past states for their current behaviour are known as **sequential systems**. Their outputs change with time, going through a sequence of states. The D flip-flop is the universal building block for such systems. It is a central feature of counters, registers and memories, all of which play a large part in the function of computers and controllers. Just as all logic systems can be made from just NAND gates, and all analogue processing systems can be made from op-amps, all sequential systems can be built with D flip-flops.

D flip-flop

The circuit symbol and timing diagram for a D flip-flop are shown in Fig. 5.17.

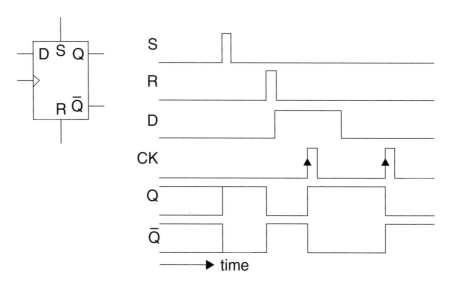

Fig 5.17 Timing diagram for a D flip-flop.

A D flip-flop has lots of terminals. Let's deal with the outputs first. There are two of them, Q and $\overline{Q}$. These outputs are **complementary**, meaning that they always have opposite states, as shown in the bottom two rows of the timing diagram.

Two of the inputs can be used to set and reset the flip-flop. So holding S high forces Q to be 1 and $\overline{Q}$ to be 0. Similarly, holding R high instead forces Q to be 0 and $\overline{Q}$ to be 1. When S and R are both low, Q remains stable at whatever it was when SR first became 00.

Edge-triggering

Of the two remaining input terminals, one has no name on its symbol. The **clock input** is represented by a triangle where the signal CK enters the flip-flop. This is because the flip-flop is **edge-triggered**. Each time that CK rises from 0 to 1, the state of D is copied to Q and then frozen. This is not the same a latch, where the state of D is copied to Q while EN is high, leaving Q frozen when EN is low. Edge-triggering not only allows for very precise timing of data capture, it also permits the construction of sequential systems with feedback paths.

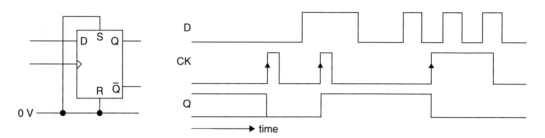

Fig 5.18 A D flip-flop can only change its state on the rising edge of clock pulses.

The timing diagram in Fig. 5.18 shows how D flip-flops are normally used for data capture. Both S and R inputs have been held low, so that the flip-flop is in one of its stable states, with Q frozen. The signal at D is set high or low, depending on what you want to store. Once D is stable, CK goes from 0 to 1 and D is instantly copied to Q. Notice that changes of D while CK is high have no effect on Q, nor does the falling edge at CK. Only **rising edges** (shown with an upwards arrow on the timing diagram) are able to unfreeze the state of Q.

Quiz referee

Fig. 5.19 shows two D flip-flops in action to decide which of two switches (Y or Z) get pressed first. Only one of the buzzers can sound at any time, indicating which switch was pressed first.

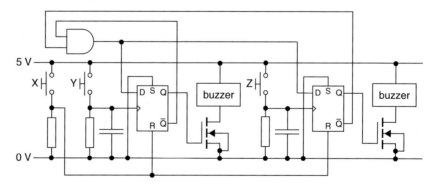

Fig 5.19 A quiz referee system.

The system is initialized by briefly pressing the X switch. This resets both flip-flops, forcing Q low and switching off both buzzers, and leaving both $\overline{Q}$ terminals high. Both inputs to the AND gate are high, so its output feeds a 1 into both D inputs. Suppose that switch Y is now pressed. The left-hand flip-flop receives a rising edge at its clock terminal, instantly copying the 1 at D to Q and lowering $\overline{Q}$ to 0. The AND gate responds in under a microsecond, lowering D to 0. If switch Z is now pressed, the 0 at D will get copied to Q, resulting in no change. In other words, as soon as one switch has been pressed, the other is locked out and has no effect.

Storing signals

Switch bounce

You may have noticed that the circuit of Fig. 5.19 on the previous page contains a couple of capacitors. These are there to ensure that pressing a switch results in only one rising edge at a clock terminal. A switch is a mechanical device. When it is pressed, two slabs of metal are slammed against each other. These slabs often bounce off each other several times before settling down. Each time the slabs separate and come together, a rising edge is generated as shown in Fig. 5.20.

More than one rising edge for each press of a switch is often bad news for systems which contain edge-triggered components. The inclusion of a capacitor in parallel with the pull-down resistor solves the difficulty. As you can see from Fig. 5.21, the capacitor charges instantly on the first contact of the metal slabs, discharging through the resistor each time the slabs bounce and lose contact. Provided that the time constant of the circuit is greater than the time between bounces (typically a few milliseconds), the result is a single rising edge.

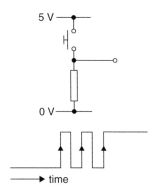

Fig 5.20 Closing a switch creates more than one rising edge.

Fig 5.21 The capacitor stops bounces of the switch resulting in rising edges.

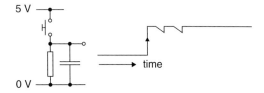

Master and slave

A D flip-flop achieves the useful trick of edge-triggering by placing a pair of latches in series as shown in Fig. 5.22. This arrangement is called a **master–slave pair**, and works because only one of them can be transparent at any one time.

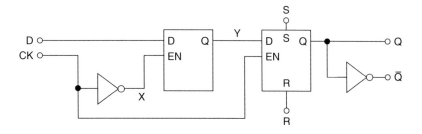

Fig 5.22 Master–slave pair of latches to make a D flip-flop.

This is how the system responds to signals which arrive at the clock terminal:

1 When CK is low the left-hand latch is transparent, so the current state of D is copied to Y; the right-hand latch is frozen, so changes at Y have no effect.

2 As CK goes high, the left-hand latch freezes and the right-hand one becomes transparent, allowing the state of Y to be copied to Q and its inverse to $\overline{Q}$.

3 While CK is high, the state of Y cannot change, so Q and $\overline{Q}$ are frozen even though the right-hand latch is transparent.

4 When CK goes low, the right-hand latch freezes the state of Q and $\overline{Q}$, so no subsequent changes of Y can affect the output of the flip-flop.

The result is a system whose output can only change while CK goes from 0 to 1. Although this requires the use of 23 NAND gates, the result is an extremely useful basic system.

OCR Electronics for AS

Registers

Arrays of D flip-flops are very good at capturing and storing binary words at a particular time in computer systems. Such arrays are known as **registers**. Fig. 5.23 shows the arrangement of D flip-flops for a four-bit register, one that can store the binary equivalent of a number between 0 and 9.

Fig 5.23 A four-bit register.

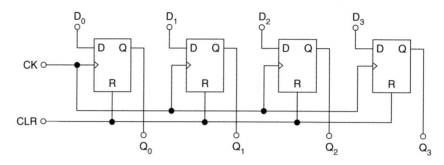

The contents of the register can be cleared by pulsing CLR high. This resets all four bits (Q_0 to Q_3) to 0. The word to be stored is then placed at the four input terminals D_0 to D_3. A rising edge at CK transfers the word to the outputs Q_0 to Q_3, where it remains frozen.

Computation

Registers are widely used in computers to hold binary words steady so that they can be processed. For example, consider the system shown in Fig. 5.24.

Fig 5.24 Four-bit word comparator.

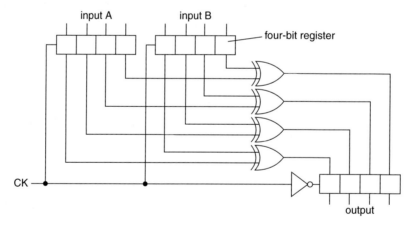

This is how it works:

1 A rising edge at CK loads two four-bit words into the pair of registers at the top.

2 While CK is high, the EOR gates compare the words bit by bit, presenting the result at the four inputs of the register at the bottom.

3 When CK goes low, the bottom register transfers the four signals at its inputs and freezes them at the output.

The use of edge-triggered devices means that this comparison of two words can happen very quickly. You just need to make sure that the rising and falling edges of CK are far apart enough for the outputs of the EOR gates to be stable. This may take up to a microsecond, allowing the system to make up to a million comparisons each second.

Storing signals

Questions

5.1 Bistables

1 The output of the bistable shown in Fig. Q5.1 can be set or reset by signals at S and R, respectively.

   (a) What is the meaning of the terms **set** and **reset**?

   (b) Write out a truth table for a single NOR gate.

   (c) Use the truth table to explain why Q can be 1 or 0 when the word SR = 00.

   (d) Complete this table for the bistable.

S	R	Q
0	0	1 or 0
1	0	
0	1	

   (e) Use a timing diagram to summarize the behaviour of the bistable.

Fig Q5.1

2 A NAND gate bistable has two active-low inputs (S, R) and one output Q.

   (a) What is the meaning of **active-low input**?

   (b) Draw a circuit for the bistable.

   (c) One of the active-low inputs can be used to set the bistable. Label that input S. Use Boolean algebra or a NAND gate truth table to justify your choice of input terminal.

OCR Electronics for AS

S	R	Q
1	1	1 or 0
1	0	
0	1	

D	EN	S	R
0	0		
0	1		
1	0		
1	1		

Fig Q5.2

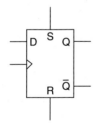

Fig Q5.3

5.2 Latches

1 Fig. Q5.2 shows the timing diagram for a latch made from a NAND gate bistable and a logic system.

  (a) Describe the behaviour of a latch.

  (b) Complete this truth table for the bistable.

  (c) Use the timing diagram to help you complete this truth table for the logic system.

  (d) Write down Boolean expressions for S and R in terms of D and EN.

  (e) Show how the logic system can be assembled from AND, OR and NOT gates.

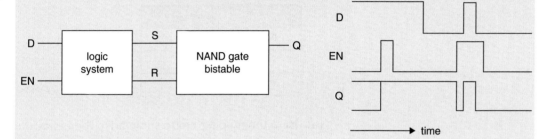

2 Latches are useful for capturing data as binary words.

  (a) Show how a number of latches can be connected to capture a four-bit word WXYZ. Label the inputs and outputs.

  (b) Describe and explain how the system can be used to capture a four-bit word.

5.3 Flip-flops

1 Fig. Q5.3 shows the input and output terminals of a D flip-flop.

  (a) Describe the effect of the S and R inputs on the outputs Q and $\bar{Q}$.

  (b) The flip-flop is **rising-edge triggered**. What does this mean?

  (c) Use a timing diagram to show how signals at D and CK affect the outputs.

2 Fig Q5.4 on the next page shows a master–slave arrangement of latches to make a D flip-flop.

  (a) Each latch can be **transparent** or **frozen**. What is the meaning of the terms in bold?

  (b) Complete the table with T (for transparent) or F (for frozen) to show how the state of each latch depends on the signal at the clock input.

clock	left-hand latch	right-hand latch
0		
1		

  (c) Explain why the signal at the input of the system can only be transferred to the output when the clock input goes from 1 to 0.

  (d) Use a timing diagram to show how the signal at the output of the system is affected by the signals at the data and clock inputs.

Storing signals

Fig Q5.4

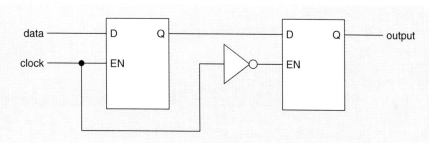

3 This question is about the processing system shown in Fig. Q5.5 below.

(a) Describe the behaviour of a four-bit register.

(b) Show how a four-bit register can be made from D flip-flops. Label the input, output and clock terminals.

(c) The words 0110 and 1101 are placed at the inputs of registers A and B. Describe and explain what happens to the system when CK is

  (i) raised high

  (ii) moved low aga

Fig Q5.5

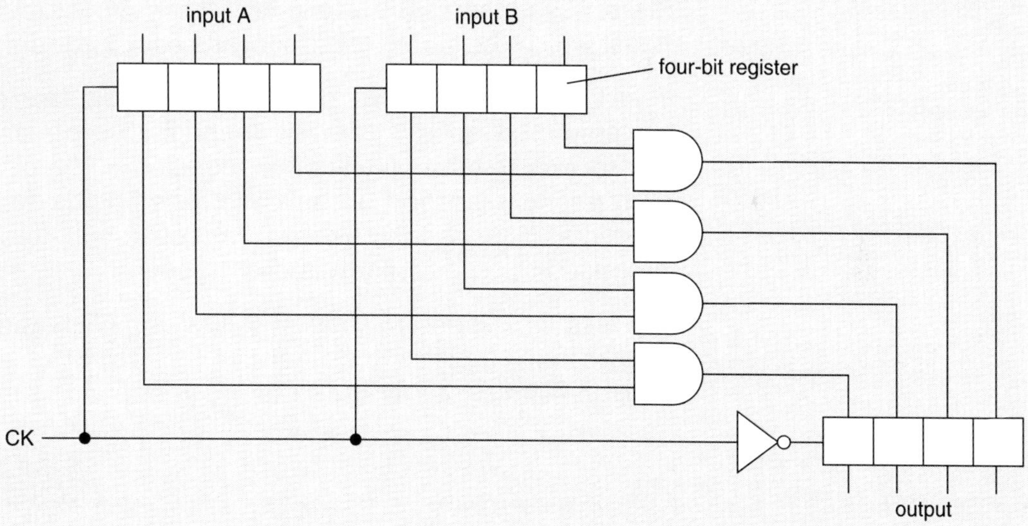

Learning summary

By the end of this chapter you should be able to:

- use a timing diagram to show the behaviour of a bistable
- use NOR and NAND gates to make bistables
- use a timing diagram to show the behaviour of a latch
- construct a latch from a bistable and logic gates
- use a timing diagram to show the behaviour of a D flip-flop
- assemble a D flip-flop from latches
- use D flip-flops to store and process binary words

CHAPTER 6

Negative feedback

6.1 Amplifiers

Every time you speak into your phone, a **microphone** transfers the information in your sound into an electrical signal. That signal passes through the processors of the phone system until it arrives at a **loudspeaker**, where the information is transferred back into sound. As far as the user is concerned, the block diagram of Fig. 6.1 summarizes the whole process. The microphone produces an electrical signal which the loudspeaker converts back into sound.

Fig 6.1 A useless audio communication system.

In fact, the system shown in Fig. 6.1 is useless. The microphone only obtains a tiny amount of energy from the sound it absorbs, far too little for the loudspeaker to make any audible sound. The 100 pW of electrical power from a microphone is far short of the 1 W needed by the loudspeaker. An **amplifier** is needed to increase the power of the signal, hopefully without altering any of the information that it contains about the sound.

This chapter is going to show you how amplifiers can be constructed from op-amps. These amplifiers have transfer characteristics which are predictable and result in faithful transfer of the sound information through the system.

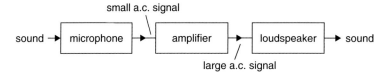

Fig 6.2 A block diagram of a useful communication system.

Microphones

A number of different technologies exist for transferring the energy of a sound wave into an electrical signal. Most of them do not need a power supply, but they do not give out much signal either. The **electret microphone** of Fig. 6.3 requires a 5 V power supply, but is therefore better able to pick up low levels of sound. It is the microphone of choice for most modern communication systems, such as mobile phones.

Fig 6.3 An electret microphone with a typical oscilloscope trace of the a.c. signal at the output.

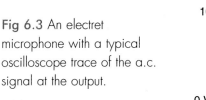
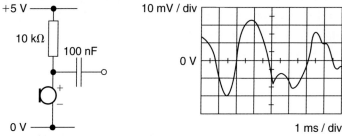

Negative feedback

The microphone has a polarity, indicated by the + and − signs at its terminals, and needs to be connected to the +5 V supply rail by a 10 kΩ resistor. As sound is absorbed by the microphone, charge moves up and down through the resistor, creating an alternating current (**a.c.**) in it. The size of this current depends on the loudness of the sound, but it is typically of the order of a few microamps. The current in the resistor results in an alternating voltage which passes through the 100 nF **coupling capacitor** to the output terminal. Fig. 6.3 on the previous page shows the trace of the output signal for a typical sound wave on the screen of an oscilloscope. Notice that it alternates between positive and negative voltages, with an average of value zero, a **peak value** of about 20 mV and **period** of a few milliseconds.

Coupling capacitor

As charge moves back and forth through the 10 kΩ resistor, it creates an a.c. signal centred on +5 V, the voltage of the top supply rail. This is not any good for amplifiers which expect the a.c. signals at their input to be centred on 0 V. So a 100 nF **coupling capacitor** creates a copy of the a.c. signal at the base of the resistor, but centred on 0 V instead of +5 V.

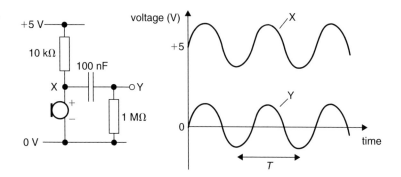

Fig 6.4 The capacitor allows the a.c. signal through but blocks the d.c.

This is illustrated in Fig. 6.4. The microphone feeds its signal into a 1 MΩ load (typical for the input of an oscilloscope). As the signal at X varies up and down in voltage, the voltage at Y varies up and down as well. However, the direct current (**d.c.**) signal of 5 V at X is blocked by the capacitor, allowing the signal at Y to be centred on 0 V.

So why does the a.c. get through but not the d.c.? Consider the time constant of the system:

$$RC = 1 \times 10^6 \times 100 \times 10^{-9} = 0.1 \text{ s or } 100 \text{ ms}$$

This means that it takes about 100 ms to change the voltage drop across the capacitor by an appreciable amount. So within a second of powering up the circuit, the 5 V d.c. will charge up the capacitor. The a.c. however changes over a much shorter timescale (much less than 100 ms), giving insufficient time for the 5 V drop across the capacitor to change. So as one plate changes voltage, the other changes with it, providing that the changes are rapid enough.

OCR Electronics for AS

Voltage gain

The job specification of an amplifier is very simple; to increase **only** the amplitude of a signal. This means that the input and output signals must have the same shape and frequency. Only the amplitude must be different, as shown in Fig. 6.5.

Fig 6.5 An ideal amplifier only changes the amplitude of a signal.

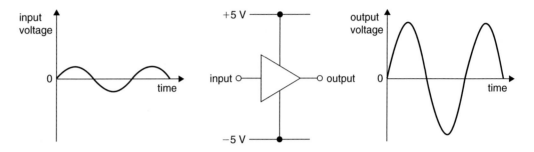

The amplifier shown in Fig. 6.5 is **ideal**. The only difference between the input and output a.c. signals is their amplitude. The peak value of the output signal is five times larger than the peak value of the input signal, giving the amplifier a **voltage gain** of five.

The voltage gain is calculated with this formula.

$$G = \frac{V_{out}}{V_{in}}$$

G is the voltage gain of the amplifier. V_{out} is the peak voltage of the signal at the output and V_{in} is the peak voltage at the input. Both of these need to be in the same units (volts, millivolts or microvolts) for the calculation to be correct. The reason why G has no units may become clear from this alternative version of the voltage gain formula.

$$V_{out} = G \times V_{in}$$

Transfer characteristic

The transfer characteristic for the ideal amplifier of Fig. 6.5 is shown in Fig. 6.6. It shows how the voltage at the output depends on the voltage at the input. There are three regions to the graph:

- The amplifier saturates at -5 V for input voltages below -1 V.
- The amplifier is linear for input voltages between -1 V and $+1$ V.
- The amplifier saturates at $+5$ V for input voltages above $+1$ V.

The saturation levels are determined by the supply rails at $+5$ V and -5 V. These limit the maximum swing of the output voltage to $+5$ V and -5 V for an ideal amplifier. The linear region is where the amplifier obeys the formula $V_{out} = G \times V_{in}$. Since G is $+5$, the amplifier can only be linear when the input voltage is between $+1$ V and -1 V, otherwise the output voltage would be above $+5$ V or below -5 V.

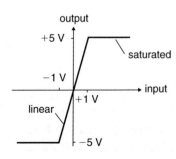

Fig 6.6 Transfer characteristic for an ideal amplifier.

Negative feedback

Distortion

Real amplifiers often **distort** the signals passing through them. This is particularly true for the amplifier circuit shown in Fig. 6.7.

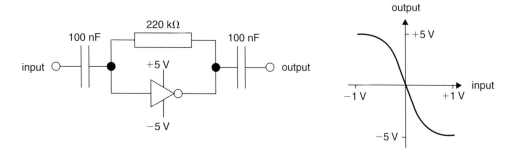

Fig 6.7 Transfer characteristic for an amplifier made from a NOT gate.

It features a NOT gate run off supply rails at +5 V and −5 V. The 220 kΩ resistor applies feedback to bias the input and output voltages to the same value (about 0 V) when there is no signal at the input. A pair of coupling capacitors allow a.c. signals in and out of the system. The transfer characteristic shows that the amplifier has a negative voltage gain of about −15 for input signals below 0.1 V, but very non-linear for signals above 0.5 V. This is shown in the oscilloscope traces of Fig. 6.8.

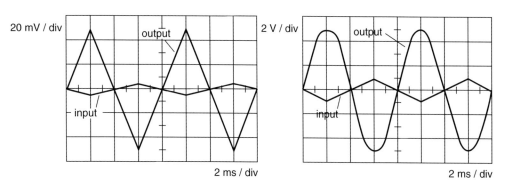

Fig 6.8 Oscilloscope traces for small and large signals at the amplifier input and output.

The test signal is a **triangle wave**. Its straight line sections and sharp edges make it particularly easy to spot changes in shape of the signal as it passes through the amplifier. As you can see from the left-hand trace, no distortion is apparent. The voltage gain can be estimated by reading off the voltages at input and output 2.0 ms after the start of the trace:

$$G = \frac{V_{out}}{V_{in}} = \frac{+50 \text{ mV}}{-4 \text{ mV}} = -13$$

The right-hand trace shows what happens when the input signal strays into the non-linear part of the transfer characteristic. The output signal is now much closer to a **sine wave** than a triangle wave, so will sound quite different. This is not a good thing. The distortion is predictable with an input signal whose peak value is +1.0 V.

$$V_{out} = G \times V_{in} = -13 \times 1.0 \text{ V} = -13 \text{V}$$

The calculated output is well below the supply rail limit of −5 V, making saturation inevitable.

OCR Electronics for AS

6.2 Voltage followers

Given the right connections, op-amps can be made into amplifiers with near-ideal behaviour. The simplest of these amplifiers, the **voltage follower**, is shown in Fig. 6.9.

The feedback loop gives the whole system a voltage gain of exactly 1. Before you find out how connecting the output and inverting input terminals achieves this, you ought to consider why an amplifier with a voltage gain of only one might be worth having.

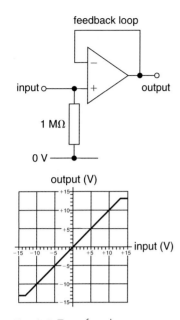

Fig 6.9 Transfer characteristic for an op-amp connected as a voltage follower.

Fig 6.10 A voltage follower provides power gain.

Power out

Consider the circuit shown in Fig. 6.10, where a sine wave of peak value 2.5 V is processed by a voltage follower before being passed to a loudspeaker.

A glance at the graph of Fig. 6.9 shows that the signal across the loudspeaker must be exactly the same as the signal at the non-inverting terminal of the op-amp. However, the power delivered to the loudspeaker is much greater than that provided by source of the sine wave signal. Let's confirm this by finding the power transferred to the loudspeaker. We need to start by finding the peak current.

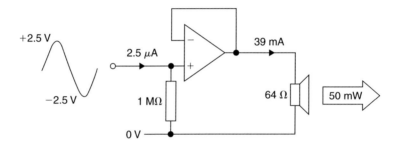

$I = ?$

$V = 2.5$ V

$R = 64\ \Omega$

$I = \dfrac{V}{R} = \dfrac{2.5}{64} = 3.9 \times 10^{-2}$ A or 39 mA

(This is about the maximum current you can extract from the output of a TL084 op-amp.) Now we can calculate the peak power of the loudspeaker.

$P = ?$

$V = 2.5$ V

$I = 3.9 \times 10^{-2}$ A

$P = IV = 3.9 \times 10^{-2} \times 2.5$

$= 9.8 \times 10^{-2}$ W or 100 mW

The average power of a sine wave signal is half the peak power, so the average power coming out of the loudspeaker is 50 mW. Enough to make a very audible sound.

Negative feedback

Power in

Now compare this with the power drawn from the signal source. Start off by calculating the peak current in the **pull-down resistor** at the non-inverting input.

$I = ?$

$V = 2.5 \text{ V}$

$R = 1 \text{ M}\Omega$

$I = \dfrac{V}{R} = \dfrac{2.5}{1 \times 10^6} = 2.5 \times 10^{-6} \text{ A or } 2.5 \text{ }\mu\text{A}$

The current in the non-inverting input is much smaller than this, less than 1 nA, so it can be safely ignored. Now for the peak power in the pull-down resistor.

$P = ?$

$V = 2.5 \text{ V}$

$I = 2.5 \times 10^{-6} \text{ A}$

$P = IV = 2.5 \times 10^{-6} \times 2.5$

$= 6.3 \times 10^{-6} \text{ W or } 6 \text{ }\mu\text{W}$

The average power drawn from the signal source is therefore only 3 μW. This is a lot smaller than the 50 mW being supplied to the loudspeaker. (The extra energy comes from the supply rails which are connected to the op-amp.)

Pull-down resistor

The power drawn from the signal source is determined by the resistance of the pull-down resistor. Why not increase the power gain of the voltage follower by getting rid of the pull-down resistor so that the current from the signal source is less than 1 nA? Well, if you do not have the pull-down resistor, the op-amp may not function correctly. Each input carries a small leakage current of about 25 pA which has to go somewhere. The 1 MΩ resistor provides a safe path to the 0 V rail for this current, for a voltage drop of only 25 μV. This makes the voltage at the output of the follower slightly different from the voltage at its input, but the difference is usually too small to be a problem.

Difference amplifier

So why does the negative feedback shown in Fig. 6.9 make an op-amp into an amplifier with a voltage gain of 1? Take a look at the transfer characteristic of an op-amp shown in Fig. 6.11.

As you can see, as soon as V_- goes above V_+, the output voltage V_{out} goes from saturating positively to saturating negatively. This behaviour can be described with this formula.

$$V_{out} = A(V_+ - V_-)$$

The quantity A is the **open-loop gain** of the op-amp, and needs to be very big. For a TL084 op-amp it is a massive 2×10^5. This means that for an op-amp not to be saturated at, say, 13 V, the voltage difference between its terminals must be less than $13 \div 2 \times 10^5 = 6.5 \times 10^{-5}$ V. This is effectively zero.

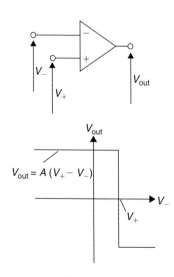

Fig 6.11 Transfer characteristic of an op-amp.

Closing the loop

Fig 6.12 The op-amp output settles to where the two graphs cross.

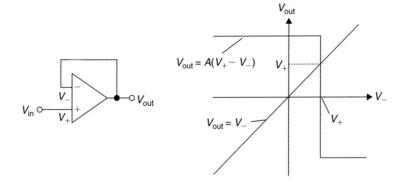

Connecting the output of an op-amp directly to its inverting input (Fig. 6.12) means that its behaviour has to be described with two formulae instead of just one.

$$V_{out} = V_-$$
$$V_{out} = A(V_+ - V_-)$$

The overall behaviour of the system can be found by eliminating V_- from the equations, as follows.

1 Replace V_- with V_{out} in the second equation.

$$V_{out} = A(V_+ - V_{out})$$

2 Eliminate the brackets.

$$V_{out} = AV_+ - AV_{out}$$

3 Gather all the terms involving V_{out} to the left.

$$V_{out} + AV_{out} = AV_+$$

4 Reintroduce brackets on the left.

$$V_{out}(1 + A) = AV_+$$

5 Finally, divide both sides by $(1 + A)$.

$$V_{out} = \frac{A}{1 + A} V_+$$

A	$\frac{A}{1+A}$
20	0.95238
200	0.99503
2000	0.99950
20 000	0.99995
200 000	0.99999

The final equation says that the voltage at the follower's output is the voltage at its input (V_+) multiplied by $A \div (1 + A)$. As you can see from the table, the value of this quantity rapidly approaches 1 as A increases. So $V_{out} = V_+$ if A is large enough. The system has a voltage gain of one.

There is another way of coming to the same conclusion. Look at the graphs of Fig. 6.12. Each graph represents one of the formulae given at the top of this page. Since the op-amp has to obey both formulae, the output must settle to a value where the graphs cross. The vertical part of the transfer characteristic occurs when $V_- = V_+$, so $V_{out} = V_+$ at the crossing point.

Negative feedback

6.3 Known gain

There is another way of looking at the effect of negative feedback on an op-amp. Take another look at the graphs of Fig. 6.12. The output of the op-amp settles to a value which keeps both of the op-amp inputs at the same voltage. This is easy to understand if you consider the op-amp formula.

$$V_{out} = A(V_+ - V_-)$$

If negative feedback makes the system into an amplifier, then V_{out} will settle somewhere between the saturation levels of the op-amp (± 13 V in the example). This means that the difference in voltage between the inverting and non-inverting inputs is going to be very small.

$$V_+ - V_- = \frac{V_{out}}{A} \leq \frac{13}{2 \times 10^5} = 6.5 \times 10^{-5} \text{ V or } 70 \text{ } \mu\text{V}$$

Non-inverting amplifier

The circuit of Fig. 6.13 has negative feedback through a pair of resistors R_t and R_b. The result is a system with a voltage gain which obeys this formula.

$$G = 1 + \frac{R_t}{R_b}$$

To see where this formula comes from, consider the current in the feedback resistors.

$$I = \frac{V}{R} = \frac{V_{out}}{R_b + R_t}$$

Now use this current to find the voltage at F.

$$V_F = IR_b = \frac{V_{out}}{R_b + R_t} R_b$$

Because of the negative feedback, the voltages at F and the input will be the same.

$$V_{in} = \frac{R_b}{R_b + R_t} V_{out}$$

Finally, rearrange to look like the voltage gain formula.

$$\frac{R_b + R_t}{R_b} = \frac{V_{out}}{V_{in}} = G = 1 + \frac{R_t}{R_b}$$

Notice that the voltage gain G of the whole system is decided by the resistor values, not the open-loop gain A. This is good, because the value of A can vary significantly from one op-amp to the next, whereas resistor values can be quite precise and reliable.

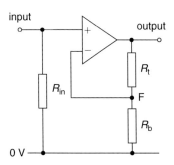

Fig 6.13 The circuit for a non-inverting amplifier.

Amplifier design

Here's the specification for an amplifier:

- Voltage gain of +15.
- Input resistance of 2.2 MΩ.
- Peak output voltage of 10 V.

You clearly want to make the non-inverting amplifier using an op-amp so that you can use the resistor values to set the gain to the value required.

$$G = 1 + \frac{R_t}{R_b}$$

$$15 = 1 + \frac{R_t}{R_b}, \text{ so } 14 = \frac{R_t}{R_b}$$

You need to deceide on a value for one of the resistors. The bottom resistor R_b will always be smaller than the top resistor R_t, so it is best to choose that one first. This is because the available current at the output of an op-amp is limited. It makes sense to have all resistors connected to the op-amp at least 1 kΩ – this limits the output current to below 13 mA, safely below the TL084 maximum limit of 40 mA. Let's go for $R_b = 3.3$ kΩ.

$$14 = \frac{R_t}{3.3 \times 10^3}, \text{ so } R_t = 3.3 \times 10^3 \times 14 = 4.6 \times 10^4 \, \Omega, \text{ about 47 kΩ}$$

The input resistance 2.2 MΩ is below the maximum safe value of 10 MΩ for a TL084 op-amp. The leakage current of 25 pA in it will result in an apparent extra d.c. input signal of 55 μV – hardly likely to be a problem if the voltage gain is only 15.

Supply rails

The peak output voltage of an amplifier can be used to set the lower limit on the voltage of the supply rails. The output of an op-amp usually struggles to get within 2 V of its supply rails, particularly if it has to source or sink appreciable amounts of current. So if you want a peak voltage of 10 V, you could get away with supply rails at ±12 V, as shown in Fig. 6.14. Notice that the amplifier is linear, provided that it does not saturate. This is because the resistance of the resistors remains fixed as the current in them varies.

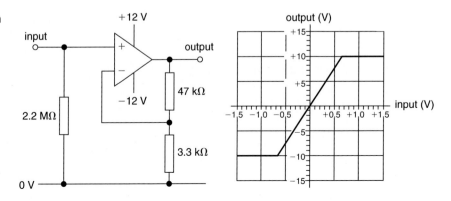

Fig 6.14 Linear amplifier with a gain of +15, with its transfer characteristic.

Negative feedback

Inverting amplifier

The amplifier in Fig. 6.15 obeys this formula.

$$G = -\frac{R_f}{R_{in}}$$

Its voltage gain is always negative, hence it is an **inverting amplifier**.

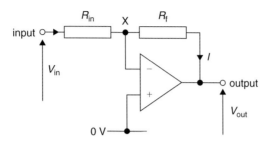

Fig 6.15 An op-amp connected as an inverting amplifier.

The feedback resistor R_f ensures that both of the op-amp input terminals have the same voltage. Since the non-inverting input is firmly held at 0 V, the inverting input will also sit at 0 V. So the point marked X in Fig. 6.15 is often called a **virtual earth** – it behaves as though it is connected to ground without being directly connected to it. Of course, this can only happen when the output terminal is not saturated, when it is free to move up and down.

Gain formula

The key to understanding the voltage gain formula is an appreciation of the current in the resistors. The current is the same in both because the current at the inverting input is negligible. Fig. 6.15 shows the direction of the current in each resistor. It should be clear why V_{in} and V_{out} must have opposite signs if X is at 0 V.

Here is the route to obtaining the voltage gain formula. Start by considering the current in the input resistor.

$$I = \frac{V}{R} = \frac{V_{in} - 0}{R_{in}} = \frac{V_{in}}{R_{in}}$$

Do the same for the current in the feedback resistor.

$$I = \frac{V}{R} = \frac{0 - V_{out}}{R_f} = -\frac{V_{out}}{R_f}$$

Now equate the two currents.

$$\frac{V_{in}}{R_{in}} = -\frac{V_{out}}{R_f}$$

Finally, rearrange to look like the voltage gain formula.

$$\frac{V_{out}}{V_{in}} = -\frac{R_f}{R_{in}}$$

As with all op-amp amplifier circuits, it is a good idea to have all resistor values between 1 kΩ and 10 MΩ.

Inverting amplifier design

Look at the transfer characteristic of Fig 6.16. How would you go about building an amplifier which behaves like this?

Fig 6.16 Transfer characteristic for an inverting amplifier.

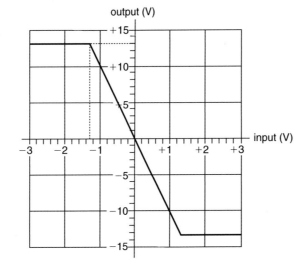

Your starting point is the sign of the voltage gain. The graph shows that the input and output voltages always have opposite signs, so the voltage gain is negative. You need an inverting amplifier. Now take any pair of values from the linear part of the graph and use them to calculate the voltage gain.

$$G = \frac{V_{out}}{V_{in}} = \frac{+13}{-1.3} = -10$$

Finally, chose a value for the input resistor (e.g. 10 kΩ) and calculate a value for the feedback resistor.

$$G = -\frac{R_f}{R_{in}} = -10 = -\frac{R_f}{10 \times 10^3}, \text{ so } R_f = 10 \times 10 \times 10^3 = 1.0 \times 10^5 \text{ Ω or } 100 \text{ kΩ}$$

Input resistor

Fig. 6.17 shows two circuits which have the transfer characteristic of Fig. 6.16. The one on the right is a better circuit as it has a much higher input resistance. This minimizes the power drawn from the source of whatever signal is being amplified – always a good thing to do.

Fig 6.17 Two circuits with the transfer characteristic of Fig. 6.16.

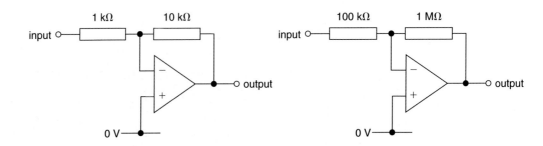

Negative feedback

6.4 Summing signals

Communication systems often need more than just microphones, amplifiers and loudspeakers. They also need **summing amplifiers**. These are devices which can combine more than one a.c. signal into a single channel. The combining is done by adding the instantaneous voltages of the signals at the various inputs (Fig. 6.18) and placing the result at the output. So, for example, if input 1 is from a lead guitar, input 2 is from a bass guitar and input 3 is from a microphone, the signal at the output contains information about the sounds from both guitars and the singer. Of course, all of this has to happen without introducing any distortion, so you will not be surprised to find out that it requires the use of an op-amp with negative feedback.

Fig 6.18 The output is the moment-by-moment sum of the signals at the three inputs.

Summing amplifier

Fig. 6.19 gives more detail about the assembly of a summing amplifier. Each of the a.c. signals to be combined (S_1, S_2 and S_3) arrives at one end of a 100 kΩ resistor, generating a current in it. The three resistors meet at the inverting input of the op-amp, so the current from each of the signals are combined here to give the current in the 33 kΩ feedback resistor. The result is that the amplifier obeys this formula.

$$-\frac{V_{out}}{R_f} = \frac{V_1}{R_1} + \frac{V_2}{R_2} + \frac{V_3}{R_3}$$

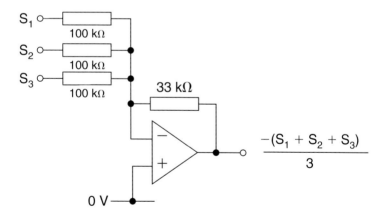

Fig 6.19 An op-amp used to combine three input signals

Insert the resistor values.

$$-\frac{V_{out}}{33 \times 10^3} = \frac{S_1}{100 \times 10^3} + \frac{S_2}{100 \times 10^3} + \frac{S_3}{100 \times 10^3}$$

Rearrange to obtain a formula for the output signal V_{out}.

$$V_{out} = -\frac{33 \times 10^3}{100 \times 10^3}(S_1 + S_2 + S_3) = -\frac{1}{3}(S_1 + S_2 + S_3)$$

So the summing amplifier adds the three signals together, divides the results by three and reverses the polarity of the resulting signal. The last operation does not make any difference to the sound of the combined signals, and the division by three ensures that the output will never be distorted by saturating at $+13$ V or -13 V (assuming that each input signal comes from an op-amp run off supply rails at ± 15 V).

Summing amplifier formula

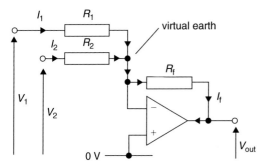

Fig 6.20 The current in the feedback resistor is the sum of the input currents.

In order to obtain the formula linking input and output signals for the summing amplifier of Fig. 6.20, you need to start by considering the currents in each input resistor, remembering that negative feedback makes the inverting input a virtual earth (0 V).

$$I_1 = \frac{V_1 - 0}{R_1} \text{ and } I_2 = \frac{V_2 - 0}{R_2}$$

These currents combine at the virtual earth to make the current in the feedback resistor.

$$I_f = I_1 + I_2 = \frac{V_1}{R_1} + \frac{V_2}{R_2}$$

The current in the feedback resistor can be used to give the voltage at the output.

$$V = IR = 0 - V_{out} = I_f R_f$$

Finally, eliminate I_f to obtain the final equation.

$$-\frac{V_{out}}{R_f} = I_f = \frac{V_1}{R_1} + \frac{V_2}{R_2}$$

The result is easily extendable to as many input signals as you wish.

Negative feedback

Digital to analogue

The summing amplifier of Fig. 6.21 does something interesting. It makes an analogue signal out of a pair of digital signals. This is the principle behind the creation of sound by computers.

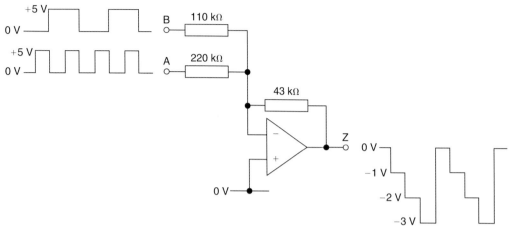

Fig 6.21 A summing amplifier making an analogue signal from digital signals.

So how does it work? Start by writing down formulae for the two input signals V_B and V_A.

$$V_B = 5B \text{ and } V_A = 5A$$

The variables A and B alternate between 1 and 0 as time goes on. Insert these formulae into the summing amplifier formula.

$$-\frac{V_Z}{43 \times 10^3} = \frac{5B}{110 \times 10^3} + \frac{5A}{220 \times 10^3}$$

Rearrange to obtain a formula for the voltage at the output Z:

$$V_Z = -43 \times 10^3 \left(\frac{5B}{110 \times 10^3} + \frac{5A}{220 \times 10^3}\right) = -(2B + A)$$

The voltage at the output for the four different output states is shown in the table.

BA	$V_Z = -(2B + A)$
00	$-(0 + 0) = 0$ V
01	$-(0 + 1) = -1$ V
10	$-(2 + 0) = -2$ V
11	$-(2 + 1) = -3$ V

Mixers

Fig. 6.22 shows a summing amplifier with potentiometers arranged as variable resistors at the inputs. This allows the value of the input resistor to be varied from 2 kΩ to 502 kΩ, altering the amount of the signal which is added to the output. The result is a system known as a **mixer**, a piece of equipment which is indispensable in a recording studio.

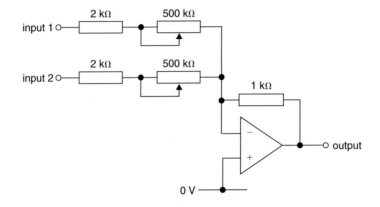

Fig 6.22 A two-input mixer with variable resistors at the inputs.

The circuit has one main disadvantage. You cannot turn either input signal off completely. This problem is solved in the circuit of Fig. 6.23, where the potentiometers allow any fraction of the input signal (from all of it to none of it) through to the fixed input resistor. Notice the choice of resistor values which ensure that the output can never saturate if the input signals stay in the range ±13 V.

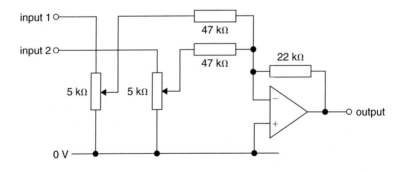

Fig 6.23 A mixer circuit which allows input signals to be turned off completely.

Negative feedback

Questions

6.1 Amplifiers

1. The amplifier in Fig. Q6.1 has a voltage gain of +6. The output terminal of the amplifier saturates at ±9 V.

   (a) Describe the behaviour of an amplifier.

   (b) A signal of amplitude 0.5 V and frequency 340 Hz is placed at the input. Calculate the amplitude and frequency of the signal at the output.

   (c) Draw a transfer characteristic for the amplifier to show how the output voltage depends on the input voltage. Mark scales on both axes.

   (d) Explain why signals at the input need to have an amplitude of less than 1.5 V if the output signal is not to be distorted.

2. This question is about the microphone circuit shown in Fig. Q6.2.

   (a) Describe the function of a microphone circuit.

   (b) Sound waves absorbed by the microphone result in a.c. currents in the 5 kΩ resistor. If the amplitude of the current is 20 μA, what is the amplitude of the a.c. voltage across the resistor?

   (c) Explain the function of the capacitor in the circuit.

   (d) A whistle results in a signal of amplitude 50 mV and frequency 2.5 kHz at the output of the circuit. Draw a voltage–time graph for two cycles of the signal. Mark scales on both axes.

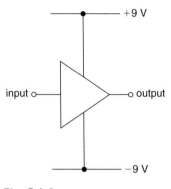

Fig Q6.1

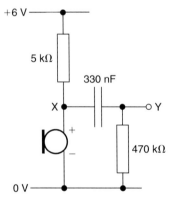

Fig Q6.2

6.2 Voltage followers

1. The output of the voltage follower of Fig. Q6.3 saturates at ±10 V.

   (a) Explain why the input and output voltages have the same value, provided that the output does not saturate.

   (b) Sketch a transfer characteristic of the circuit to show how the output voltage depends on the input voltage as it is swept from −12 V to +12 V.

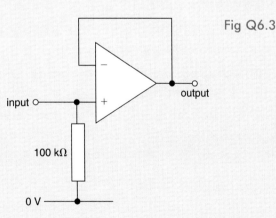

Fig Q6.3

2 A loudspeaker of resistance 16 Ω is connected to the output of the circuit of Fig. Q6.3. An a.c. signal of amplitude 5 V is applied to the input terminal.

   (a) Calculate the current in the 100 kΩ input resistor.

   (b) Calculate the peak power delivered to the follower by the signal source.

   (c) Calculate the peak power delivered to the loudspeaker by the op-amp.

   (d) Calculate the power gain of the voltage follower.

6.3 Known gain

Assume that the outputs all of the op-amps in these questions saturate at ±13 V.

1 A non-inverting amplifier is required to have these characteristics:

   - A voltage gain of +5.
   - An input resistance of 47 kΩ.

   (a) Draw a circuit diagram for the amplifier. Show all component values and justify them with calculations.

   (b) Sketch a transfer characteristic for the amplifier to show how the output voltage depends on the input voltage as it is swept from −15 V to +15 V.

2 An inverting amplifier is required to have these characteristics:

   - A voltage gain of −10.
   - An input resistance of 22 kΩ.

   (a) Draw a circuit diagram for the amplifier. Show all component values and justify them with calculations.

   (b) Sketch a transfer characteristic for the amplifier to show how the output voltage depends on the input voltage as it is swept from −15 V to +15 V.

3 Amplifiers made from op-amps with negative feedback are linear, with predictable voltage gains, provided that the amplitude of the input signal is not too large.

   (a) How is negative feedback arranged in these amplifiers?

   (b) What is the effect of negative feedback?

   (c) What is the meaning of the term **linear**, and why is it important for an amplifier to have this property?

   (d) Explain how to calculate the voltage gain of amplifiers based on op-amps with negative feedback.

   (e) Explain why op-amp amplifiers are not linear if the input signal gets too large.

Negative feedback

6.4 Summing signals

Assume that the outputs all of the op-amps in these questions saturate at ±13 V.

1. The circuit of Fig. Q6.4 combines the signals at its inputs and outputs the result at Z.

   (a) State the voltage at point X in the circuit. Give a reason for your answer.

   (b) Calculate the currents in the two 100 kΩ resistors.

   (c) Explain why the current in the 200 kΩ resistor is 30 μA.

   (d) Calculate the voltage at the output Z.

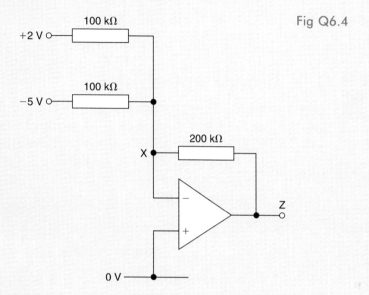

Fig Q6.4

2. Here is the specification for a three-input mixer system.
   - Each input signal is an a.c. voltage.
   - The maximum amplitude from each source is 10 V.
   - The amount of each signal mixed into the output can be varied from 100 per cent to 0 per cent
   - The maximum amplitude of the output signal is 10 V.

   Draw a circuit diagram for the mixer system, using potentiometers and a summing amplifier. Show all component values and justify their values with calculations.

Learning summary

By the end of this chapter you should be able to:

- use a microphone to transfer sound into an electrical signal
- calculate the gain of an amplifier from its input and output signals
- connect an op-amp to make a voltage follower
- use resistors to make an op-amp into a non-inverting amplifier
- use resistors to make an op-amp into an inverting amplifier
- use an op-amp to make a summing amplifier

CHAPTER 7

Counting pulses

pulse number	D	C	B	A
0	0	0	0	0
1	0	0	0	1
2	0	0	1	0
3	0	0	1	1
4	0	1	0	0
5	0	1	0	1
6	0	1	1	0
7	0	1	1	1
8	1	0	0	0
9	1	0	0	1

Fig 7.1 A four-bit counter.

7.1 Binary counters

Take a look at the component featured in Fig. 7.1. It is a **binary counter**. As pulses enter the input marked CK, the pattern of 1s and 0s at the outputs A, B, C and D changes. The word DCBA shows how many pulses have entered CK since the last time the system was reset by pulsing R high, as shown in this **pulse table**.

The pulse number can be calculated from the output word DCBA with this formula.

$$\text{Number} = 8D + 4C + 2B + A$$

where D, C, B and A are 1 or 0. This type of coding is known as **binary-coded decimal** (or **BCD**), and is internationally recognized as the representation of the numbers from 0 to 9 with a four-bit word.

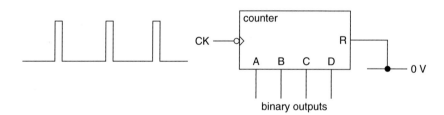

One-bit counter

Since a binary counter has to remember how many pulses have arrived at CK, it will come as no surprise that it contains a number of D flip-flops. Each D flip-flop has to be connected as shown in Fig. 7.2 on the next page, with D connected to $\overline{Q}$. This means that on every rising edge at the clock input of the flip-flop, the state of $\overline{Q}$ is copied to Q. In other words, Q changes state on each falling edge at CK.

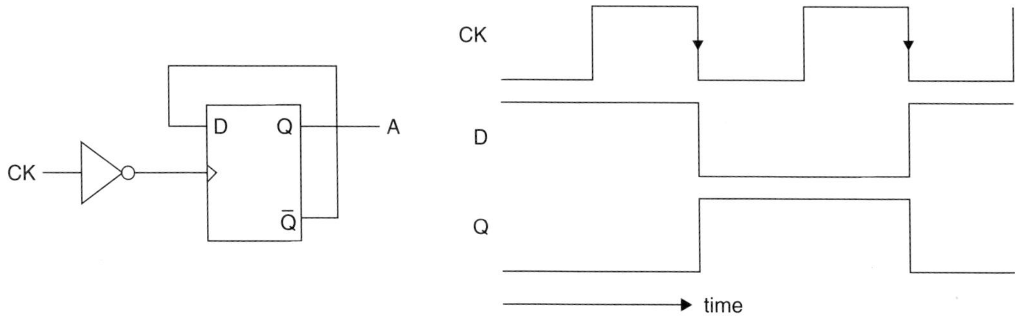

Counting pulses

Fig 7.2 Timing diagram for a one-bit counter.

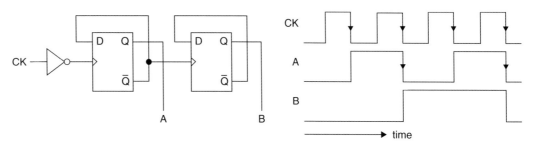

pulse number	A
0	0
1	1
2	0
3	1

Take a look at the timing diagram of Fig. 7.2. The top row shows a square wave entering the CK terminal. This is inverted by the NOT gate, making falling edges at CK produce rising edges at the clock terminal of the D flip-flop. So at each falling edge at CK, Q changes state. This can be summarized with this pulse table.

Two-bit counter

Binary counters with any number of outputs, not just four, can be made by cascading one-bit counters in series. The inverted output of each counter provides the input to the next one as shown in Fig. 7.3.

Fig 7.3 Timing diagram for a two-bit counter.

pulse number	B	A
0	0	0
1	0	1
2	1	0
3	1	1
4	0	0
5	0	1

The timing diagram shows that A changes state on every falling edge entering CK. Similarly, B changes state on every falling edge at A. After four pulses at CK the system ends up back where it started, so a two-bit counter can only code for the numbers 0, 1, 2 and 3.

Four-bit counter

Fig. 7.4 shows how the four-bit counter of Fig. 7.1 can be assembled from a chain of D flip-flops. As the output of each flip-flop falls, a rising edge is fed into the clock terminal of the next flip-flop in the chain. This ensures that the system counts up in binary. The reset terminals of the four flip-flops are connected together so that pulling R high resets DCBA to 0000. (Notice that the S terminals of the individual flip-flops have not been shown, meaning that you can assume they are permanently connected to 0 V.)

Fig 7.4 A four-bit counter made from a chain of D flip-flops.

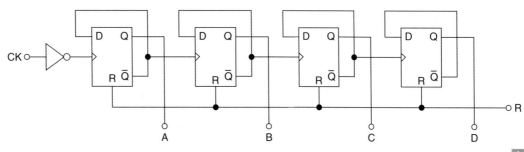

OCR Electronics for AS

Maximum count

The number of bits in the output word of a counter determines how far it can count before it resets to zero. The number of different states for a word of N bits is given by this formula.

$$\text{number of states} = 2^N$$

So a three-bit counter can have $2^3 = 8$ different states. Since the output of a counter usually starts at zero, this means that a three-bit counter can count up to seven pulses before it resets to zero. This is shown in this pulse table.

Notice that the output word for the eighth pulse is the same as it was before any pulses arrived, namely CBA = 000.

pulse number	C	B	A
0	0	0	0
1	0	0	1
2	0	1	0
3	0	1	1
4	1	0	0
5	1	0	1
6	1	1	0
7	1	1	1
8	0	0	0

One-to-five counter

Counters do not always have to start at zero or count up to their maximum limit before resetting. For example, Fig. 7.5 shows a counter which counts from one through to five, thanks to the use of an AND gate which uses C and B to control R or S for the flip-flops.

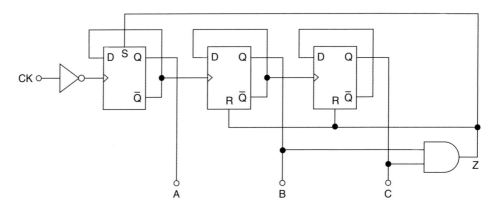

Fig 7.5 A one-to-five counter.

The connections of the output of the AND gate determine the starting point for the counter. Each time that Z is high, the first flip-flop in the chain is set and the last two are reset. This makes the output word of the counter CBA = 001, the BCD equivalent of one. So the counter starts counting at one, not zero.

The upper limit of the count is determined by the connections of the AND gate inputs. The output of the AND gate is represented by Z = C.B, so it only goes high when CBA = 110. In less than a microsecond of this happening, CBA = 001 and Z returns to zero. As the table on the next page shows, the system counts from one to five before returning to one.

Counting pulses

pulse number	C	B	A	count
0	0	0	1	1
1	0	1	0	2
2	0	1	1	3
3	1	0	0	4
4	1	0	1	5
5	0	0	1	1
6	0	1	0	2
7	0	1	1	3
8	1	0	0	4

The timing diagram of Fig. 7.6 shows how the states of C, B, A and Z change as the count proceeds. Notice that as soon as Z goes high, it goes low again as CBA changes from 110 (six) to 001 (one).

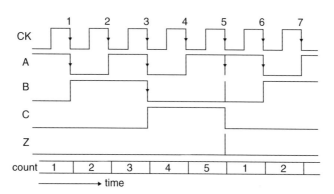

Fig 7.6 Timing diagram for the circuit of Fig.7.5.

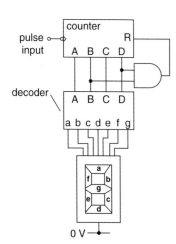

Fig 7.7 Decimal counter with decoder and seven-segment display.

Decimal counter

An AND gate can convert a binary counter into a decimal one, provided that its inputs are connected to B and D as shown in Fig. 7.7. As the system counts pulses, the first time that B and D are both high is when DCBA = 1010, the BCD equivalent of ten. So the output of the counter cycles from DCBA = 0000 (zero) to DCBA = 1001 (nine) as pulses enter the input terminal.

Decimal display

The output of a decimal counter is a four-bit word DCBA which encodes in BCD a number between zero and nine. Fig. 7.7 shows how a **decoder** can be used to display the number on a **seven-segment display**. The latter is a set of LEDs arranged in parallel (Fig. 7.8) so that they can display all of the numbers between zero and nine. The decoder is just a logic system which converts each four-bit word DCBA into the seven-bit word gfedcba required to display the BCD equivalent of DCBA. For example, when DCBA = 0011, the decoder feeds gfedcba = 1001111 into the LEDs to display the number three.

Fig 7.8 Connection details for a seven-segment LED display.

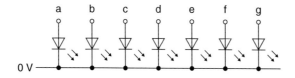

Cascading counters

Fig 7.9 This system counts in decimal from zero to ninety nine.

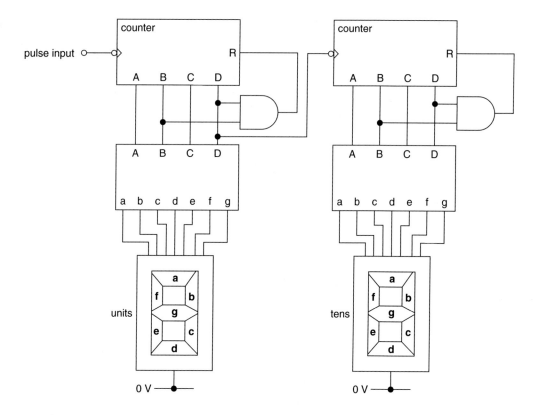

The decimal counter is a useful building block for making much larger counters. If one can count from zero to nine, then two can count from zero to ninety nine, as shown in Fig. 7.9. The first counter in the chain counts the pulses at the input, recording the result on the left-hand display. Every time that the display changes from 9 back to 0, the D output of the first counter falls from 1 to 0 (Fig. 7.10). This falling edge is counted by the second counter, so the right-hand display records the number of tens of pulses at the system input. (The pulse at R is too short to be counted reliably.) Of course, the system runs out of capacity on the hundredth pulse and displays 00 instead, but that can be easily solved by connecting a third decimal counter on the right.

Fig 7.10 Timing diagram for a single decimal counter.

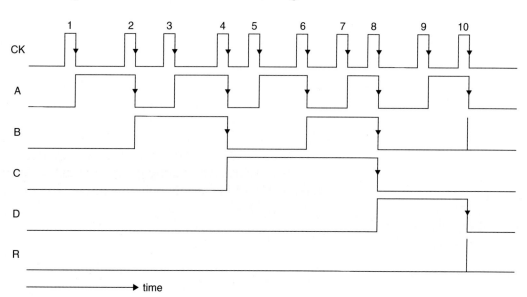

Counting pulses

7.2 Clocks

The block diagram of an electronic clock is shown in Fig. 7.11. The input is an oscillator which produces a supply of pulses at a known and steady rate, such as one per second. These are processed by a counter system which determines the numbers of seconds, minutes, hours and days which have happened since the system was last reset. The output block is some sort of display which conveys the information stored in the counter to human observers – it could use LEDs, but is more likely to use a liquid crystal display (LCD).

Fig 7.11 Block diagram for an electronic clock.

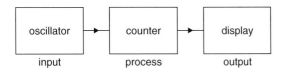

Crystal oscillators

All clocks have to contain something which tells them how long a second is. All good electronic clocks use a **crystal** to determine the length of a second. This is a small slab of quartz inside a capacitor which has a natural frequency of oscillation, typically 32 768 Hz. When made part of a suitable circuit (such as the one shown in Fig. 7.12), the crystal can force it to oscillate producing a square wave with a frequency of 32 768 Hz which is stable to within at least 10 parts in a million. (This is equivalent to losing or gaining time by at most 1 second per day.)

Fig 7.12 Crystal oscillator.

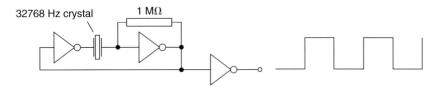

Why 32 768 Hz? Well, it is exactly 2^{15} Hz, which makes it easy to use a chain of counters to generate a signal which has a frequency of 1 Hz (one pulse per second), as shown in Fig. 7.13. Consider any flip-flop in the chain. One cycle of its output ($0 \to 1 \to 0$) requires two pulses at its clock input ($0 \to 1 \to 0 \to 1 \to 0$). So two cycles of the square wave at the input results in only one cycle at the output. Each flip-flop in the counter therefore outputs a signal which has half the frequency of the signal at its input. A square wave signal at 2^{15} Hz at the input of a chain of 15 flip-flops results in a signal at 1 Hz out of the last flip-flop.

Fig 7.13 Counter chain to convert 32 768 Hz into 1 Hz.

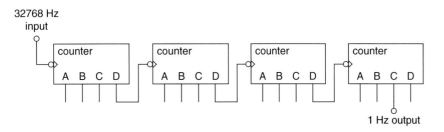

Clock sub-systems

The crystal oscillator and its fifteen-bit counter chain are only one of several **sub-systems** in an electronic clock. As you can see from Fig. 7.14, there is a chain of three separate sub-systems which display the seconds, minutes and hours of the time. Each takes in a signal from the previous sub-system and produces the signal for the next one in the chain, as well as decoding the signal for display.

Fig 7.14 Sub-systems of a clock's processing block.

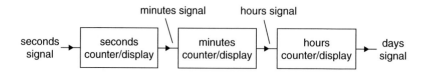

Counting seconds

Fig. 7.15 shows the details of the sub-system which processes and displays the time in seconds. One falling edge per second arrives at the clock terminal of the left-hand binary counter. Every tenth pulse causes the counter to reset, and the falling edge at its D terminal provides a clock pulse for the right-hand counter. The sixth of these pulses makes that counter reset, firing off one falling edge per minute (60 s) into the next sub-system.

Fig 7.15 Seconds counter and display for a clock.

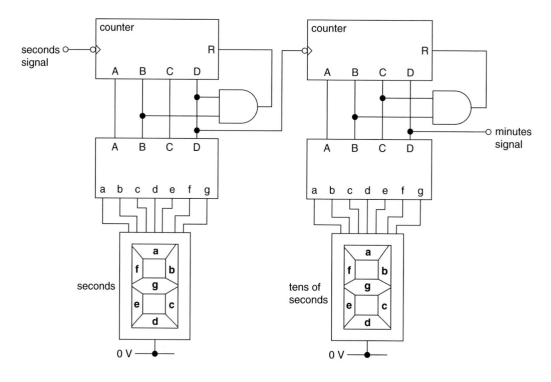

Counting pulses

Counting hours

The sub-system which deals with hours is more complicated, as it has to count from 00 to 23 before resetting. As you can see from Fig. 7.16, the left-hand counter resets on every tenth pulse at its clock input. The right-hand counter only resets when its count is two and the left-hand counter has recorded four pulses. The OR gate allows the reset signal for the right-hand counter to also reset the left-hand one, so that the display shows 00 every twenty-four hours.

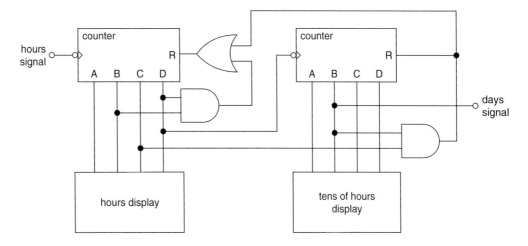

Fig 7.16 Hours counter for a 24-hour clock.

Stopwatch

A stopwatch is just a clock which starts or stops counting when a switch is pressed. The block diagram for one is shown in Fig. 7.17. It incorporates a latch to allow a single switch to start, stop and reset the system. The switch feeds start/stop pulses into the one-bit counter, so each pressing of the switch changes the state of its output. A high output lets pulses from the oscillator through to the counter, as well as enabling the latch, allowing the display to show the time passing. The next pressing of the switch makes the output of the one-bit counter low. This immediately stops pulses from the oscillator getting to the counter. It also makes the latch freeze at the same instant. Shortly afterwards (thanks to the **propagation delay** of the signal through the NOT gate), the counter is reset to zero, ready to start counting again. The result is a system which counts up from zero when a switch is pressed and released for the first time, only to have the display freeze when the switch is pressed for the second time.

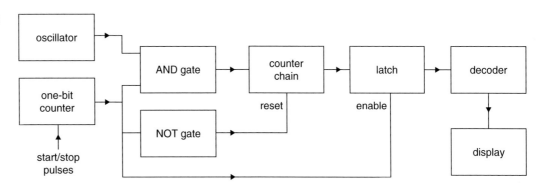

Fig 7.17 Block diagram for a stopwatch.

Frequency division

A clock is a special example of a system which performs **frequency division**. It uses a long chain of counters to convert a 1 Hz signal into output signals at 1/60 Hz (one per minute), 1/3600 Hz (one per hour) and 1/86 400 Hz (one per day). A much simpler system that performs frequency division is shown in Fig. 7.18. For every two falling edges at X, there is only one at Y, so the frequency at Y is half that at X. Similarly, the frequency at Z is a quarter that of the signal at X – four falling edges at X are required to make one at Z. The table shows how the frequency at the output of a chain of counters depends on N, the number of flip-flops in the chain.

Fig 7.18 A simple frequency divider.

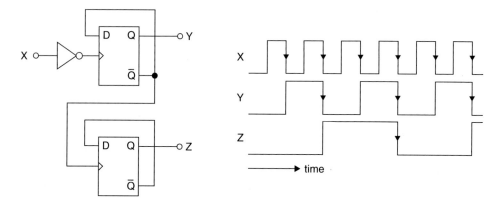

number of flip-flops, N	frequency division
1	$f \rightarrow f \div 2$
2	$f \rightarrow f \div 4$
3	$f \rightarrow f \div 8$
4	$f \rightarrow f \div 16$

Division by three

Frequency division by 2^N is straightforward. Just feed a square wave of frequency f into the input of an N-bit binary counter and out comes a square wave of frequency $f \div 2^N$. But what if the divisor you need is not given by 2^N? You then need a system like the one shown in Fig. 7.19, which divides the frequency of the signal at its input by three.

Fig 7.19 The signal at the input is three times the frequency of the signal at the output.

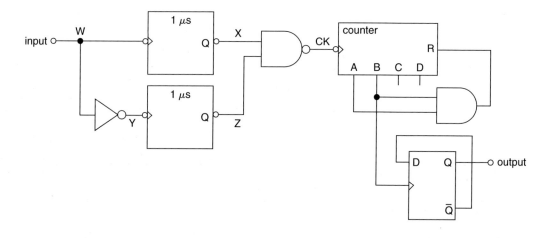

Counting pulses

Fig 7.20 Timing diagram for the frequency doubling sub-system.

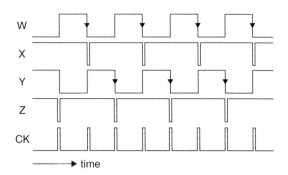

The timing diagram of Fig. 7.20 shows the processing done by the two monostables and the NAND gate. This sub-system produces two short pulses in each cycle of the square signal W, one pulse for each rising and falling edge. These pulses are combined by the NAND gate to make a signal which has two falling edges in each cycle of the input signal at W. This provides the clock signal CK for the counter. The counter is reset when BA = 11, the binary equivalent of three. Each time that B rises it triggers the one-bit counter into changing state as shown in the timing diagram of Fig. 7.21. The result is a square wave signal at Q, with one falling edge for every three at W. So the output of the system is a copy of the input, but with one-third of the frequency.

Fig 7.21 There are three falling edges at W for every one at Q.

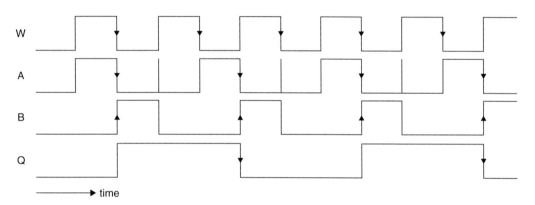

Division by anything

By now, it should be obvious how to obtain frequency division by any number you like. Generate a signal at double the original frequency f, feed it into a binary counter which is reset after N counts, feed the most significant bit of the counter outputs into a one-bit counter, and you are left with a square wave at frequency $f \div N$. The block diagram for such a general-purpose frequency division system is shown in Fig. 7.22.

Fig 7.22 Block diagram for a general-purpose frequency division system.

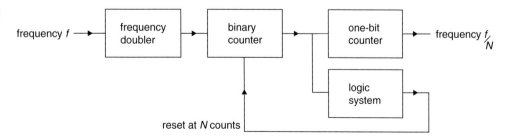

OCR Electronics for AS

7.3 Continuous sequencers

A set of traffic lights is a good example of a continuous sequencer. The system has three outputs (red, amber and green lamps) which go on and off according to a fixed pattern which is repeated over and over again. This section is going to show you how to design such systems.

Clock, counter and logic

The block diagram for a continuous sequencer is shown in Fig. 7.23. The clock produces pulses for the counter. Each pulse at the input of the counter moves its outputs onto the next state in the standard binary sequence. The counter outputs are processed by the logic system to generate an appropriate reset signal for the counter as well as the required sequence of digital signals at the output.

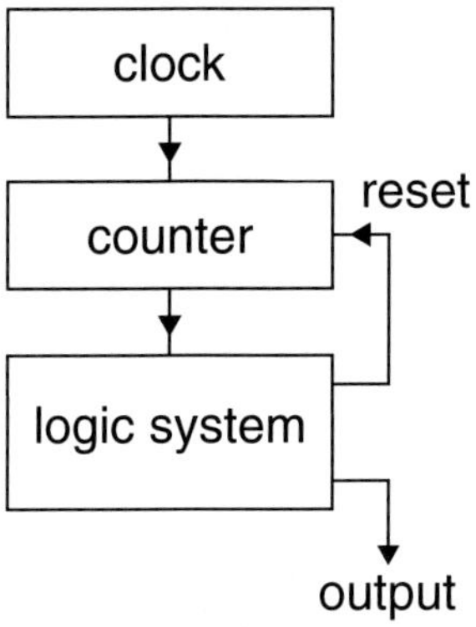

Fig 7.23 Block diagram for a continuous sequencer.

Specification

Here is the specification of a small continuous sequencer system:

- It has three LEDs (red, yellow and green) as the output.
- Each LED only glows for 0.5 s at a time.
- The LEDs glow one after the other, in the order red, yellow, green, red, yellow, ...

The first step in designing this system is to produce a table to show the states of the outputs at each step of the sequence.

state	red	yellow	green
0	on	off	off
1	off	on	off
2	off	off	on
3	on	off	off
4	off	on	off

Notice that the numbering of the states starts at zero, not one. This is because a binary counter is going to keep track of the states, and it starts at zero after every reset.

State table

This **state table** shows how the outputs of the logic system (R, X, Y and Z) will be related to the counter outputs (B and A), where X, Y and Z control the LEDs via MOSFET drivers (Fig. 7.24).

state	B	A	R	X	Y	Z
0	0	0	0	1	0	0
1	0	1	0	0	1	0
2	1	0	0	0	0	1
3	1	1	1	–	–	–

Counting pulses

Fig 7.24 The LEDs come on one after the other, for half a second at a time.

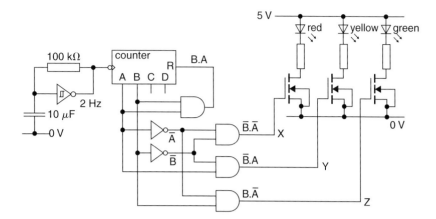

Inspection of the state table gives you four Boolean algebra expressions for the logic system.

$$R = B.A$$

$$X = \overline{B}.\overline{A}$$

$$Y = \overline{B}.A$$

$$Z = B.\overline{A}$$

These are easily implemented with basic logic gates as shown in Fig. 7.24. Notice that you do not have to worry about the states of X, Y and Z in state 3 because it is only there for less than a microsecond as shown in the timing diagram in Fig. 7.25 below.

Clock speed

The timing of each state does not have to be very precise, so a relaxation oscillator has been used as the clock for Fig. 7.24. Each falling edge from the clock moves the system from one state to the next, at the end of every half a second. You can verify for yourself that the formula $T = 0.5RC$ shows that the component values (100 kΩ, 10 μF) do indeed give a period of 0.5 s, corresponding to a frequency of 2 Hz. Of course, a 32 768 Hz crystal oscillator and a chain of fourteen one-bit counters would do the same job, but with much more hardware.

Fig 7.25 Timing diagram for Fig. 7.24.

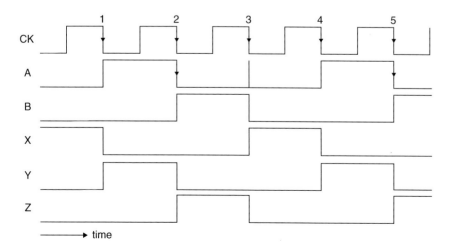

7.4 One-shot sequencers

A one-shot sequencer is a bit like a monostable whose output can be tailored to your requirements. Feed a rising edge into it and a sequence of words appears at the output. It has not only to generate a sequence, but it also has to know when to start and stop producing it. It therefore is quite a complicated system. A simple example is shown in Fig. 7.26.

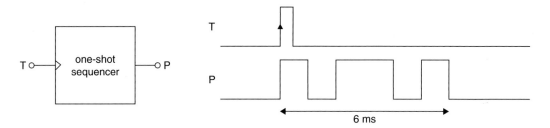

Fig 7.26 Timing diagram for a one-shot sequencer.

Circuit operation

The circuit diagram for a one-shot sequencer is shown in Fig. 7.27. This is how it operates. You might want to keep an eye on the timing diagram of Fig. 7.28 on the next page as you study each of these stages:

1. A rising edge at T sets the flip-flop.

2. This enables the 1 kHz oscillator and also resets the counter via a monostable.

3. Falling edges at CK enter the counter at intervals of 1 ms.

4. As the counter outputs change, the logic system alters P to generate the sequence.

5. After 6 ms, the logic system pulls Z low to reset the flip-flop.

6. The oscillator stops pulsing, so the counter outputs freeze.

7. The system waits for the next rising edge at T.

Notice how the flip-flop controls the flow of pulses from the oscillator, and that it can only do this if its reset terminal is initially low, so that it can recognize rising edges at its clock terminal. Since the resting state of the monostable controlled by Z is high, this start-up condition is guaranteed.

Fig 7.27 Circuit diagram for a one-shot sequencer.

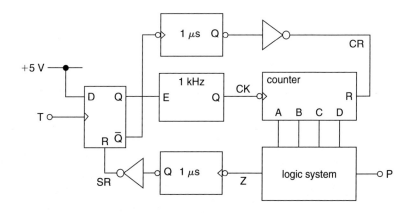

Counting pulses

Fig 7.28 Sequence of signals inside the one-shot sequencer.

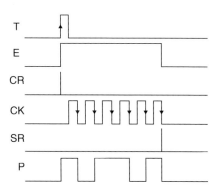

Oscillator details

Notice that the system requires the use of an oscillator which produces its first falling edge 1 ms after the counter reset signal CR has been pulsed high, as well as producing falling edges at intervals of 1 ms. You might be tempted to use a relaxation oscillator like the one shown in Fig. 7.29, with component values which give the system a frequency of 1 kHz. Unfortunately, the first cycle of CK produced when EN goes high is much longer than subsequent pulses, as shown in the timing diagram.

Fig 7.29 Poor choice of oscillator for the one-shot sequencer.

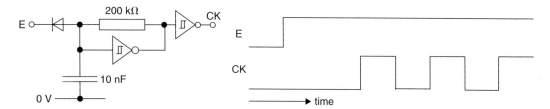

A better choice of oscillator is shown in Fig. 7.30. It still uses a relaxation oscillator, but it is free-running at 16 kHz. A four-bit counter performs frequency division, so that CK has a frequency of $16 \div 2^4 = 1$ kHz, as required. When the enable input E is low, the counter is reset, with DCBA = 0000. As soon as E goes high, R goes low and the counter is able to respond to falling edges at its clock input. Each state lasts for $1 \div 16 = 0.063$ ms. At the sixteenth pulse, D falls low for the first time. Depending on the state of the oscillator at the instant that R went low, the time delay between E going high and CK going low is between 0.94 ms and 1.00 ms. Subsequent falling edges at CK happen at intervals of 1.00 ms.

If this level of jitter in the first state of the sequence is unacceptable, you need to use a longer chain of flip-flops in the counter chain and a correspondingly faster oscillator. For example, a 64 kHz oscillator and a six-bit counter reduces the jitter to an acceptable 0.02 ms. Of course, if you need precise timing, a crystal oscillator is a much better choice than a relaxation oscillator. Not only are the trip points slightly different for each Schmitt trigger NOT gate, the values of the capacitor and resistor are likely to be a few per cent different from their stated values.

Fig 7.30 Better choice of oscillator for the one-shot sequencer.

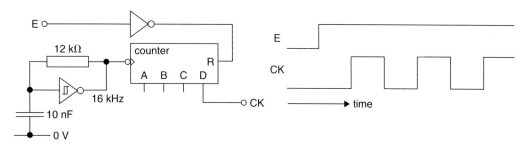

OCR Electronics for AS

Logic system

The logic system for the one-shot sequencer of Fig. 7.27 on page 112 has to obey this pulse table.

state	C	B	A	P	Z
0	0	0	0	1	1
1	0	0	1	0	1
2	0	1	0	1	1
3	0	1	1	1	1
4	1	0	0	0	1
5	1	0	1	1	1
6	1	1	0	0	0

Notice that state 6 is the resting state of the system, where it is left immediately after the flip-flop has been reset and the oscillator has been turned off. So Z must go low as the system passes from state 5 to state 6, and P must be low. Inspection of the truth table gives these two expressions for Z and P.

$$\overline{Z} = C.B$$

$$P = \overline{C}.\overline{B}.\overline{A} + \overline{C}.B.\overline{A} + \overline{C}.B.A + C.\overline{B}.A$$

They can both be manipulated to give simpler expressions.

$$Z = \overline{C} + \overline{B}$$

$$P = \overline{C}.\overline{B}.\overline{A} + \overline{C}.B + C.\overline{B}.A$$

Fig. 7.31 shows how to implement this with NAND gates.

Fig 7.31 Implementing the logic system with NAND gates.

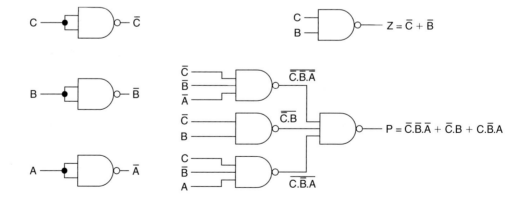

Counting pulses

Read-only memory

Designing an efficient logic system for a large sequencer, whether one-shot or continuous, can be difficult. It certainly involves skill and lots of practice to use the minimum number of logic gates. The larger the system gets, the worse the problem becomes – sequencers which run through hundreds of states and have dozens of outputs are not unusual. An alternative method, which uses more hardware, but is very easy to apply is shown in Fig. 7.32. It uses a form of **read-only memory (ROM)** which can be scaled up to any size you like for very little design effort.

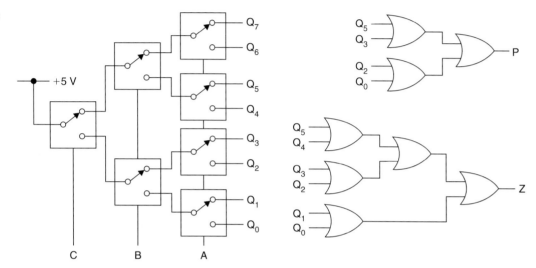

Fig 7.32 A read-only memory implementation of the logic system.

On the left, a series of demultiplexers are nested to make a one-of-eight **address decoder**. The state of the address CBA input to the system determines which one of the decoder outputs Q_0 to Q_7 are high. So as CBA counts up in binary, the decoder outputs go high one after the other, starting with Q_0. The three nested OR gates on the right combine signals from the four decoder outputs which go high during the sequence to generate P. The other nested OR gates generate Z in a similar fashion. The use of a ROM to implement a logic system goes straight from the truth table to connections between decoder outputs and nested OR gate inputs – no need for Boolean algebra at all!

Race Hazards

The implementation of Z in Fig 7.32 is risky. It could lead to a **race hazard** if a demultiplexer output goes low before the next one goes high. This would result in a momentary drop in the voltage at Z, triggering the monostable prematurely. It would be far safer to use a NOT gate connected to Q_6 instead.

OCR Electronics for AS

Questions

7.1 Binary counters

1. A one-bit counter which is triggered by falling edges can be made from a NOT gate and a D flip-flop.

   (a) Show how the counter can be assembled from the two components.

   (b) Draw a timing diagram to describe the behaviour of the one-bit counter.

   (c) Explain the behaviour of the system in terms of the properties of the flip-flop.

   (d) Add a second flip-flop to make a two-bit counter.

   (e) Draw a pulse table to show the sequence of output states as pulses enter the clock terminal of the two-bit counter.

2. Fig. Q7.1 shows a three-bit binary counter.

   (a) Explain why three of these counters will be required to count up to 100 clock pulses.

   (b) Show how the three counters should be connected to make a nine-bit counter.

   (c) Show how an AND gate can be used to reset the nine-bit counter every 100 clock pulses. Justify your connections of the AND gate inputs.

3. A counter system is required to display this sequence as pulses enter its clock terminal:

   two, three, four, two, three, four, . . .

   (a) Explain why the counter will require three D-flip-flops.

   (b) Draw up a pulse table to show the sequence of output states of the counter as pulses enter its clock terminal.

   (c) Show how the flip-flops and logic gates can be used to assemble the counter.

   (d) Show how a decoder and seven-segment LED can be used to display the counter output in decimal.

   (e) Describe the function of the decoder. Write out a truth table for it to display the numbers two, three and four.

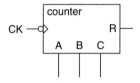

Fig Q7.1

Counting pulses

7.2 Clocks

1 Fig. Q7.2 is a block diagram for a simple electronic clock.

(a) The oscillator uses a crystal to run at 1.024 kHz. Suggest why a relaxation oscillator is **not** used.

(b) How many one-bit counters will be required to reduce the frequency of the clock output to 1.000 Hz?

(c) The output of the first counter is one falling edge every minute. Show how the counter can be assembled from four-bit counters and logic gates. Explain the operation of the counter.

(d) The clock displays hours from 0 to 12. Show how the third counter and its associated hours display can be assembled from four-bit counters, logic gates, decoders and seven-segment displays.

Fig Q7.2

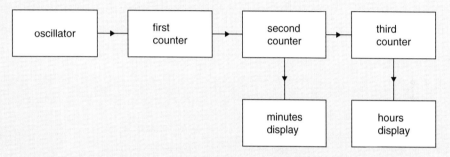

2 The system shown in Fig. Q7.3 is a frequency divider.

(a) The monostables, NOT gate and AND gate are assembled to make a frequency doubler. Draw a timing diagram for the signals at the input, L, M, N and P.

(b) Explain why the frequency of the signal at the input is five times that of the frequency at the output.

Fig Q7.3

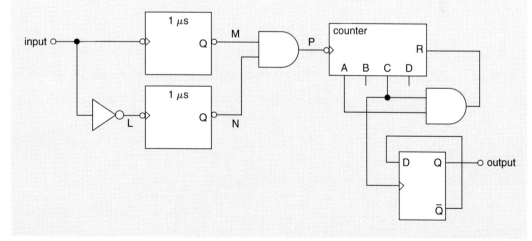

OCR Electronics for AS

7.3 Continuous sequencers

1 Fig. Q7.4 is the block diagram for a system which produces this continuous sequence of outputs on three LEDs, L, M and N:

state	L	M	N
0	on	off	off
1	on	on	off
2	off	on	off
3	off	on	on
4	off	off	on

(a) Each cycle of the sequence contains the five states shown, and lasts for ten seconds. What is the frequency of the clock? Show how it can be assembled from a relaxation oscillator. Show all component values and justify them.

(b) Explain why the system has to use a three-bit binary counter.

(c) Show how an AND gate can be used to provide the reset signal.

(d) Design a suitable logic system. Show all of the steps in your design.

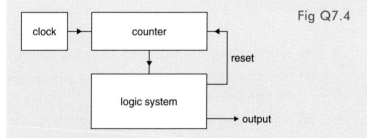

Fig Q7.4

2 This question is about the sequencer circuit in Fig. Q7.5.

(a) Calculate the frequency of the oscillator.

(b) How many states are there in the sequence? How long does each state last?

(c) Write out a pulse table for the system, showing how the signals at X, Y and Z are related to the signals at C, B and A.

(d) Use a timing diagram to show the signals at the outputs over one cycle of the sequence.

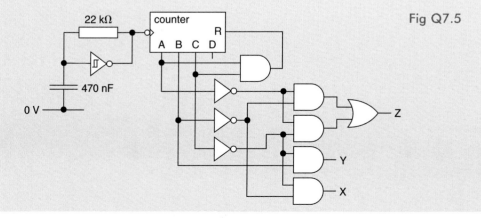

Fig Q7.5

Counting pulses

7.4 One-shot sequencers

1 A rising edge at T triggers the system in Fig. Q7.6 into producing a series of pulses at P.

(a) Before the system is triggered, the flip-flop is reset. What effect does this have on the relaxation oscillator and the binary counter?

(b) Explain why pulses appear at P when a rising edge arrives at T.

(c) Explain the function of the AND gate in the system.

(d) Explain the behaviour of the system by drawing a timing diagram for signals at T, X, P, Y and Z.

(e) Adapt the system so that eleven pulses appear at P for each rising edge at T.

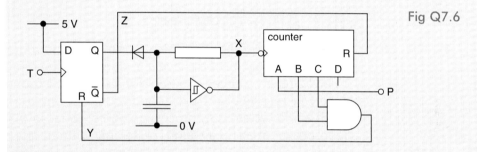

Fig Q7.6

Learning summary

By the end of this chapter you should be able to:

- use D flip-flops to make binary counters
- use logic gates to reset counters after a number of clock pulses
- represent the transfer characteristics of a binary counter with a timing diagram
- use binary counters to design clocks with decimal displays
- make systems which generate continuous sequences of digital signals
- make systems which generate a single sequence of digital signals

CHAPTER 8

Amplifying audio

8.1 Audio systems

Audio signals fed into loudspeakers produce a sound which can be heard by humans, so they need to have a frequency in the range 20 Hz to 20 kHz. There are many different sources of audio signals – the microphone, compact disc, radio tuner, TV receiver, MP3 player and mobile phone to name but a few. The function of an audio system is to process an audio signal so that it can be delivered to a loudspeaker, making a sound, preferably without any distortion or noise. In practice, this is quite difficult to achieve without careful attention to detail.

Block diagram

A useful audio system can be broken up into several different blocks, as shown in Fig. 8.1.

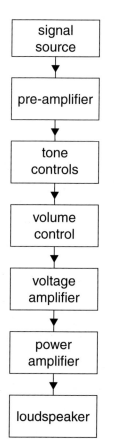

Fig 8.1 Block diagram for an audio system.

Each block in the system has its own particular function:

- The **signal source** outputs an audio signal. This is an a.c. signal whose frequency lies in the range of 20 Hz to 20 kHz, and whose amplitude is typically a few millivolts.

- The **pre-amplifier** boosts the amplitude of the audio signal so that it is well above the noise levels which the other blocks may add to it. (It is not essential to know this for the AS qualification.)

- The **tone control** adjusts the frequency balance of the signal, boosting some ranges and cutting others, compensating for the imperfect transfer characteristics of the microphone, loudspeaker and human ear.

- The **volume control** is a variable component which passes a fraction of the signal (anything between all and nothing) from the tone controls to the voltage amplifier.

- The **voltage amplifier** boosts the amplitude of the signal, typically from millivolts to volts.

- The **power amplifier** passes the signal to the loudspeaker, providing the large current required.

- The **loudspeaker** transfers the information in the audio signal as a pattern of sound waves.

Amplifying audio

Impedance matching

The block diagram of Fig. 8.1 suggests that you can take a modular approach to the design of an audio system. This entails designing and testing each block as a separate module before connecting them together to make the whole system. However, you cannot deal with each block in isolation without running the risk of poor **impedance matching**, leading to loss of signal as it transfers between blocks. As you will find out below, each signal comes out of a block through an **output impedance** and enters the **input impedance** of the next block. You need to have these two quantities in the correct ratio to persuade a high fraction of the signal to pass from one block to another.

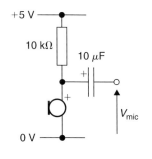

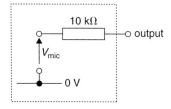

Fig 8.2 Equivalent circuit for a microphone.

Output impedance

The electret microphone at the top of Fig. 8.2 is a typical source of audio signals. The signal at its output terminal is an a.c. signal with an amplitude of a few millivolts and a frequency between 20 Hz and 20 kHz. The signal behaves as though it has to pass through a 10 kΩ resistor before it leaves the output terminal. In other words, the microphone's **output impedance** is 10 kΩ. This is shown in the **equivalent circuit** at the bottom of Fig. 8.2.

Input impedance

Now consider the next block along, the pre-amplifier. Suppose you choose the inverting amplifier on the left-hand side of Fig. 8.3. This could be a good choice because it has a large voltage gain which is also negative.

$G = ?$

$R_f = 220$ kΩ

$R_{in} = 4.7$ kΩ

$$G = -\frac{R_f}{R_{in}} = -\frac{220 \times 10^3}{4.7 \times 10^3} = -47$$

As you will find out below, a negative gain improves the stability of the whole system. The equivalent circuit for the amplifier is shown on the right-hand side of Fig. 8.3. Signals entering the amplifier source current into the **input impedance** of 4.7 kΩ, and the amplifier feeds out a signal which is −47 times the signal at the input terminal. The output impedance of an op-amp with negative feedback is much less than 1 Ω, so it has been replaced with a wire in Fig. 8.3.

Fig 8.3 Equivalent circuit for an inverting amplifier.

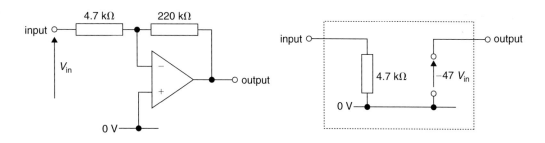

OCR Electronics for AS

Poor matching

Fig. 8.4 shows the microphone module connected to the pre-amplifier module, with the equivalent circuit for the part of the circuit where the microphone meets the amplifier. As you can see, any signal from the microphone is placed across both the 10 kΩ and 4.7 kΩ impedances, but only the signal across the 4.7 kΩ input impedance is passed to the amplifier.

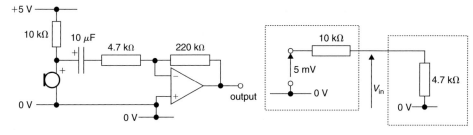

Fig 8.4 Connecting the signal source to the pre-amplifier.

A calculation should make this clearer. Suppose that the instantaneous microphone signal is 5 mV. Start by calculating the current in the two impedances.

$I = ?$

$V = 5$ mV

$I = \dfrac{V}{R} = \dfrac{5 \times 10^{-3}}{14.7 \times 10^{3}} = 3.4 \times 10^{-7}$ A or 340 nA

$R = 10 + 4.7 = 14.7$ kΩ

The signal at the input of the amplifier is the voltage across just the 4.7 kΩ input impedance.

$V = ?$

$I = 3.4 \times 10^{-7}$ A

$V = IR = 3.4 \times 10^{-7} \times 4.7 \times 10^{3}$

$R = 4.7$ kΩ

$= 1.6 \times 10^{-3}$ V or 1.6 mV

So a 5 mV signal at the microphone results in only 1.6 mV at the input of the amplifier. The voltage divider formed by the output and input impedances behaves like an amplifier with a voltage gain of much less than one.

$G = ?$

$V_{out} = 1.6$ mV

$G = \dfrac{V_{out}}{V_{in}} = \dfrac{1.6}{5} = 0.32$

$V_{in} = 5$ mV

So 68 per cent of the signal fails to get from the microphone to the amplifier. This amount of signal loss is a bad thing and can be avoided by obeying the following golden rule of **impedance matching**.

Signals should come out of a low impedance into a high impedance.

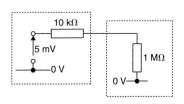

Fig 8.5 A non-inverting amplifier makes a better pre-amplifier.

Good matching

Fig. 8.5 shows a much better choice for the pre-amplifier. The non-inverting amplifier has an input impedance of 1 MΩ, one hundred times larger than the output impedance of the microphone. You can verify for yourself, by repeating the calculations above, that only 1 per cent of the microphone signal fails to get to the amplifier. This is quite acceptable.

Amplifying audio

Instability

You may have noticed that the voltage gain of the pre-amplifier in Fig. 8.5 is only a modest +11.

$G = ?$

$R_t = 10 \text{ k}\Omega$

$R_b = 1 \text{ k}\Omega$

$$G = 1 + \frac{R_t}{R_b} = 1 + \frac{10 \times 10^3}{1 \times 10^3} = 1 + 10 = +11$$

The inverting amplifier of Fig. 8.4 had a much larger voltage gain (−47). Why not have a non-inverting amplifier with a gain of +47? The reason has to do with the **stability** of the whole audio system. Amplifiers with a large positive gain are quite likely to become unstable and go into oscillation, making the whole system howl – the loudspeaker produces a loud whistle or hum. This is because the wires at the input and output of an amplifier behave like a capacitor, usually with a tiny capacitance, and a.c. signals pass easily through capacitors. So a signal at the output of the amplifier can pass through it repeatedly, getting larger every time. An inverting amplifier, however, is immune from this problem as any stray signals from the output emerge amplified with a change of sign, reducing the amplitude of the signal at the output. So inverting amplifiers are usually stable, even with voltage gains of −50 (about as large as you can get out of a TL084 op-amp over the audio range of 20 Hz to 20 kHz).

Tone control

The tone controls of an audio system alter the frequency balance of the signal from the pre-amplifier. They allow the user of the system to selectively boost or cut signals with high or low frequencies. There are a number of reasons why you may want to do this:

- The loudspeaker may not have the same efficiency across the whole audio frequency range.

- A recorded signal may not be played back at the original volume. This makes a difference because the human ear has difficulty detecting low frequency sound when it has a low volume.

- The signal source may be a digital-to-analogue converter effectively limiting the highest frequency in the audio signal.

The graph of Fig. 8.6 is the transfer characteristic of a tone control when it is set to **flat response**, with a constant gain of 1 over the whole audio range. Notice that the graph has a **logarithmic scale** on both axes, with values increasing by a factor of ten from one grid line to the next, instead of just going up by one. This is done for two reasons:

- It allows data which covers a large range of values to be compressed into a small space.

- It reflects the response of the human ear.

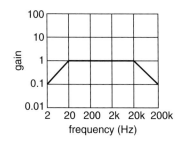

Fig 8.6 Transfer characteristic of a tone control with a flat response for a.f. signals.

OCR Electronics for AS

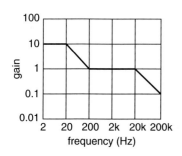

Fig 8.7 Transfer characteristic for bass boost.

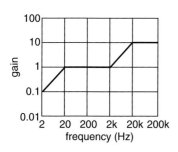

Fig 8.8 Transfer characteristic for treble boost.

Fig 8.9 Block diagram for a tone control.

Fig 8.10 Partial block and circuit diagrams for an audio system.

Bass boost

Recorded sound that is played back at a lower volume than the original always sounds wrong. It does not have enough **bass** – the low frequencies do not have enough volume. Setting the tone control to **bass boost** can compensate for this. A transfer characteristic for this setting is shown in Fig. 8.7. Signals below 200 Hz are progressively boosted, with signals at 20 Hz being ten times larger than they were before. The boosting stops at 20 Hz, to prevent overloading of the voltage amplifier by signals that cannot be heard.

Treble boost

Setting the tone control to **treble boost** can compensate for the poor performance of loudspeakers at high frequencies. A transfer characteristic (or **gain–frequency graph**) for a tone control set to treble boost is shown in Fig. 8.8. Signals above 2 kHz are progressively boosted until 20 kHz, where the boosting stops.

Switched control

The block diagram for a tone control is shown in Fig. 8.9. The signal from the pre-amplifier is switched to one of three **filters** (bass boost, flat response or treble boost) depending on the setting of the control switch. A summing amplifier combines the signals from the filters to generate the output for the next stage, the volume control.

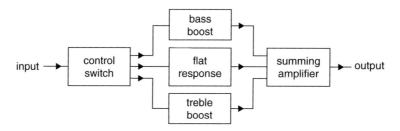

Volume control

The potentiometer in Fig. 8.10 acts as the volume control for the audio system. The setting of the **wiper** along the 1 kΩ resistance **track** determines what fraction of the signal at the input gets passed to the voltage amplifier. For example, if the wiper is halfway along the track, then only half of the signal is passed on. The 10 μF coupling capacitor blocks any d.c. signals generated in the previous stages. (A d.c. signal which is processed by the voltage amplifier would result in unnecessary heating of the loudspeaker and might overload the power amplifier.)

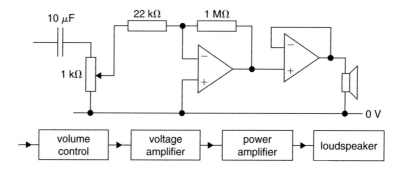

Amplifying audio

Power amplifier

The power amplifier shown in Fig. 8.10 is a voltage follower, made by imposing negative feedback on an op-amp to give it a gain of +1. The op-amp output needs to be able to handle large currents. For example, the L272M integrated circuit contains a pair of op-amps whose outputs can handle up to 1 A, allowing you to pump a mean power of up to 4 W into an 8 Ω loudspeaker. Here is the calculation. Start off by calculating the peak voltage across the loudspeaker.

$V = ?$

$I = 1 \text{ A}$ $\qquad V = IR = 1 \times 8 = 8 \text{ V}$

$R = 8 \text{ Ω}$

Now you can calculate the peak power delivered to the loudspeaker.

$P = ?$

$V = 8 \text{ V}$ $\qquad P = IV = 1 \times 8 = 8 \text{ W}$

$I = 1 \text{ A}$

The mean power of an a.c. signal is half the peak power, so the mean power delivered to the loudspeaker is $8 \div 2 = 4$ W.

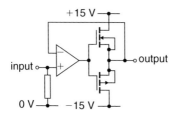

Fig 8.11 A pair of MOSFETs inside the feedback loop increases the current at the output.

Supply rails

As well as delivering power to the loudspeaker, the power amplifier also generates a lot of heat in itself. This is not a good thing. Although you cannot avoid this waste heating, you can minimize it by not having the supply rails further apart than necessary. Since the output of the L272M op-amp in Fig. 8.10 need not go above 8 V, it does not need supply rails above +10 V or below −10 V.

More power

If you need more current for higher power output, then a pair of MOSFETs inside the feedback loop can do the trick (Fig. 8.11). The arrangement of two power op-amps shown in Fig. 8.12 also increases the maximum power delivered to the speaker without needing to increase the voltage of the supply rails. The top op-amp is configured as a voltage follower, with a gain of +1, whereas the bottom op-amp is configured as an inverting amplifier with a gain of −1. The result is that the signals at X and Y have the same shape and amplitude but opposite polarity. So when X is at +8 V, Y is −8 V and the voltage drop across the speaker is 16 V. If you use L272 op-amps you can deliver a mean power of 16 W into an 8 Ω loudspeaker. That's a lot of sound!

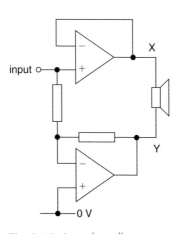

Fig 8.12 A push–pull output arrangement to increase the speaker power.

Power transfer

An op-amp with negative feedback has almost zero output impedance. This is not a good idea for the last stage of an audio system, where long wires will transfer the signal to a loudspeaker. Wires can get mangled and short-circuits often happen. Connecting the output of a power op-amp straight to 0 V in this way will not be good for it, so it is good practice to add some sort of short-circuit protection. One way of doing this is to add an extra resistor between the op-amp output and the output terminal to limit the current to a safe value. The 5 Ω resistor in Fig. 8.13 performs this function.

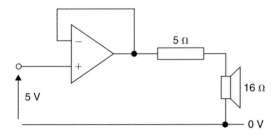

Fig 8.13 The output impedance of the power amplifier reduces the power transfer to the loudspeaker.

The system now has an output impedance of 5 Ω. This will affect the power transfer to the loudspeaker. To find out how much power is transferred, start off by calculating the current at the output. Assume that the peak signal at the op-amp output is 5 V and the loudspeaker impedance is 16 Ω.

$I = ?$

$V = 5$ V

$R = 5 + 16 = 21$ Ω

$I = \dfrac{V}{R} = \dfrac{5}{21} = 0.24$ A

Now find the peak voltage drop across the loudspeaker.

$V = ?$

$I = 0.24$ A

$R = 16$ Ω

$V = IR = 0.24 \times 16 = 3.8$ V

Finally, calculate the peak power delivered to the loudspeaker.

$P = ?$

$V = 3.8$ V

$I = 0.24$ A

$P = IV = 0.24 \times 3.8 = 0.91$ W

You can check for yourself that without the extra 5 Ω resistor, the peak power delivered to the loudspeaker is 1.6 W. This means that the inclusion of the 5 Ω protection resistor results in only about 60 per cent of the power from the op-amp being delivered to the loudspeaker. The remaining 40 per cent becomes waste heat in the protection resistor.

Amplifying audio

8.2 Filters

This section shows you how to use combinations of capacitors and resistors to cut out signals in a given frequency range. Resistors have the same behaviour, whatever the frequency of the signal across them. They always obey the formula $V = IR$, however rapidly the voltage changes sign. Capacitors, however, are different. Their use as coupling capacitors to block d.c. signals and transmit a.c. signals, as shown in Fig. 8.14, suggests that their behaviour depends on the frequency of the signal across them.

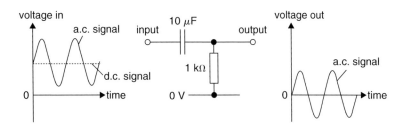

Fig 8.14 Capacitors block d.c. signals but transmit a.c. ones.

Coupling capacitors

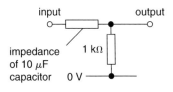

Fig 8.15 The capacitor is equivalent to a resistor whose resistance depends on frequency.

Take a look at the equivalent circuit of Fig. 8.15, where the 10 μF coupling capacitor in Fig. 8.14 has been replaced with a resistor. The resistance of that resistor (the **impedance** of the capacitor) depends on the frequency of the a.c. signal across it. It determines how much of an a.c. signal present at the input is transferred to the output. So for a.c. signals in the audio range, the impedance must be much less than the 1 kΩ resistor, allowing the a.c. signal through. On the other hand, a.c. signals of lower frequency are blocked, suggesting that the impedance must be much greater than 1 kΩ. These two situations are contrasted in Fig. 8.16 and Fig. 8.17.

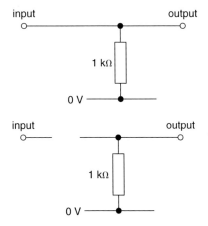

Fig 8.16 The impedance is small for signals above 20 Hz.

Fig 8.17 The impedance is very big for signals below 20 Hz.

Impedance

The effect of a coupling capacitor in Fig. 8.14 suggests that its impedance behaves as follows:

- Infinite for signals which do not alternate at all (zero frequency).
- Finite for signals which do alternate.
- Decreases as the signal frequency increases.

RC network

Fig 8.18 Testing a filter with a square wave signal.

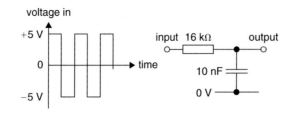

The RC network shown in Fig. 8.18 blocks the transfer of high-frequency a.c. signals. This can be shown by feeding a square wave signal into the input, and using an oscilloscope to look at the output waveform for a variety of different frequencies.

As you can see from Fig. 8.19, when the frequency is 100 Hz there is plenty of time for the capacitor to charge and discharge between rising and falling edges of the input signal. This is because the time constant is small compared with the period of 10 ms.

$\tau = ?$

$R = 16\ k\Omega$ $\qquad \tau = RC = 16 \times 10^3 \times 10 \times 10^{-9}$

$C = 10\ nF$ $\qquad\qquad = 1.6 \times 10^{-4}$ s or 0.16 ms

The situation is quite different when the frequency is increased to 10 kHz. The time constant of 0.16 ms is longer than the 50 μs between rising and falling edges of the input signal, resulting in only a small change of voltage across the capacitor. As you can see from Fig. 8.19, the signal at the output is correspondingly small.

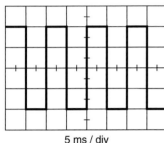

Fig 8.19 Oscilloscope traces for signals at 100 Hz and 10 kHz.

Break frequency

Fig. 8.20 shows the signal at the output of the RC network when the frequency is 1 kHz. The time between rising and falling edges is now 500 μs, about three times the time constant of 160 μs. This gives the capacitor time to almost fully charge and discharge each time, so most of the input signal makes it through to the output. This situation requires the frequency of the signal to equal the **break frequency**, given by this formula.

$$f_0 = \frac{1}{2\pi RC}$$

You can verify this with a calculation.

$f_0 = ?$

$R = 16\ k\Omega \qquad f_0 = \dfrac{1}{2\pi RC} = \dfrac{1}{2\pi \times 16 \times 10^3 \times 10 \times 100^{-9}} = 995$ Hz

$C = 10\ nF$

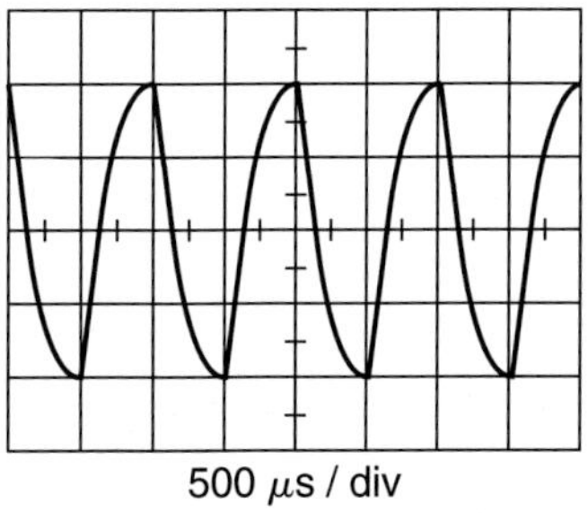

The impedance of the capacitor is the same as the resistance of the resistor at the break frequency, so the circuit has quite different behaviour above and below that frequency.

Fig 8.20 Oscilloscope trace for the signal at the break frequency of 1 kHz.

Amplifying audio

Treble-cut filter

The behaviour of an RC network as a filter is summarized in the gain–frequency graph of Fig. 8.21. Below the break frequency of 1 kHz, the impedance of the capacitor is much greater than 16 kΩ, so all of the signal gets through. Therefore the low-frequency gain is 1. However, as the signal frequency rises above 1 kHz, the impedance of the capacitor becomes smaller than 16 kΩ, resulting in a loss of signal at the output. In fact, only 10 per cent of the signal gets through to the output when the signal frequency is ten times the break frequency. So the high-frequency gain drops rapidly with increasing frequency.

Fig 8.21 Gain–frequency graph for a treble-cut filter.

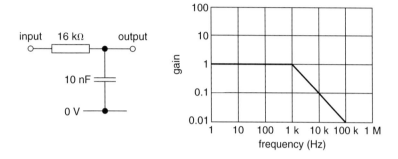

Bass-cut filter

So how do you set about analysing the behaviour of the filter circuit shown in Fig. 8.22?

Fig 8.22 Gain–frequency graph for a bass-cut filter.

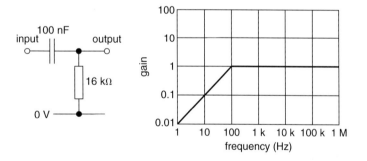

Start off by calculating the break frequency.

$f_0 = ?$

$R = 16$ kΩ

$C = 100$ nF

$$f_0 = \frac{1}{2\pi RC} = \frac{1}{2\pi \times 16 \times 10^3 \times 100 \times 100^{-9}} = 100 \text{ Hz}$$

Now think about the effect of the capacitor impedance below this frequency. It will be much larger than 16 kΩ, blocking transfer of signal through to the output. So the gain drops with frequency below 100 Hz. At frequencies above 100 Hz, the impedance will be much less than 16 kΩ, allowing most of the signal to arrive at the output. So above 100 Hz, the gain is flat at +1. Below 100 Hz, it drops by a factor of ten for each tenfold drop in frequency.

Passive filters

Both types of filter that you have met so far are known as **passive filters**. This means that they do not need a separate power supply. This can be good thing, leading to simple circuits, but does have a couple of disadvantages:

- They can be affected by the input impedance of whatever is connected to the output.
- They can only cut a signal, never boost it.

Consider the first disadvantage. As soon as the filter output is connected to another stage, the input impedance of that stage may affect the filter. This is shown in Fig. 8.23.

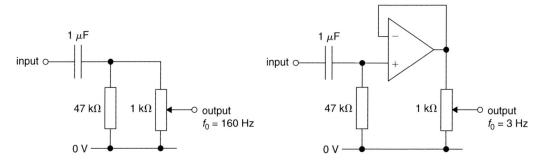

Fig 8.23 The voltage follower isolates the passive filter from the next stage.

The RC network on the left should have a break frequency of 3 Hz.

$f_0 = ?$

$R = 47\ \text{k}\Omega$

$C = 1\ \mu\text{F}$

$$f_0 = \frac{1}{2\pi RC} = \frac{1}{2\pi \times 47 \times 10^3 \times 1 \times 10^{-6}} = 3\ \text{Hz}$$

The performance of the filter is ruined by its connection to the volume control – the effective resistance of the RC network becomes the 1 kΩ of the potentiometer instead of the 47 kΩ of the resistor, raising the break frequency to 160 Hz. The problem can be solved by putting a voltage follower immediately after the filter, copying the voltage of the signal across the resistor without drawing much current from it. This is shown on the right of Fig. 8.23. Of course, the use of an op-amp requires a power supply, making the system into an **active filter**. And if you have to use an op-amp, you might as well boost the signal at the same time to drown out noise introduced by the components of the RC network.

Active treble cut

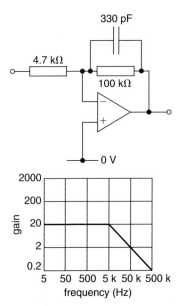

Fig 8.24 Gain–frequency graph for an active treble-cut filter.

Active filters are designed around inverting amplifiers as shown in Fig. 8.24. Although the voltage gain of this circuit is always negative, the sign of the gain can be safely ignored for a filter because it does not alter the sound of the signal. As you can see from the graph, the gain is flat at 20 for frequencies below 5 kHz. Above 5 kHz, the gain drops by ten for each tenfold increase in frequency.

Amplifying audio

Filter analysis

So how do you set about drawing a gain–frequency graph for the filter shown in Fig. 8.24? The starting point is a calculation of the break frequency for the RC network.

$f_0 = ?$

$R = 100 \text{ k}\Omega$

$C = 330 \text{ pF}$

$$f_0 = \frac{1}{2\pi RC} = \frac{1}{2\pi \times 100 \times 10^3 \times 330 \times 10^{-12}} = 5 \text{ kHz}$$

This tells you that at 5 kHz the impedance of the capacitor is 100 kΩ, the same as the resistor in parallel with it.

Now consider the situation for low frequency signals at only 500 Hz, ten times lower than the break frequency. The impedance of the capacitor will now be $100 \times 10 = 1000 \text{ k}\Omega$, much larger than the 100 kΩ resistor. So most of the charge flow in the feedback network will be through the resistor instead of the capacitor, allowing you to ignore the capacitor completely. This is shown at the top of Fig. 8.25.

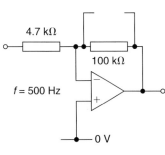

So what is the voltage gain of the system at 500 Hz?

$G = ?$

$R_f = 100 \text{ k}\Omega$

$R_{in} = 4.7 \text{ k}\Omega$

$$G = \frac{R_f}{R_{in}} = \frac{100 \times 10^3}{4.7 \times 10^3} = 21$$

So the gain–frequency graph is flat at about 20 below the break frequency of 5 kHz.

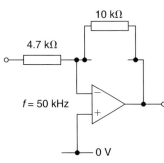

The circuit at the bottom of Fig 8.25 shows the effective circuit for signals at 50 kHz, ten times higher than the break frequency. The impedance of the capacitor is now ten times smaller than 100 kΩ, allowing most of the charge flow in the feedback network to ignore the resistor and pass through the capacitor instead. The feedback resistor is now only 10 kΩ.

Fig 8.25 Effective circuits for the filter at 500 Hz and 50 kHz.

$G = ?$

$R_f = 10 \text{ k}\Omega$

$R_{in} = 4.7 \text{ k}\Omega$

$$G = \frac{R_f}{R_{in}} = \frac{10 \times 10^3}{4.7 \times 10^3} = 2.1$$

So the gain drops to 2.1 at 50 kHz. You can check for yourself that it becomes 0.21 at 500 kHz, when the impedance of the capacitor drops to only 1 kΩ.

Two lines

The gain–frequency graph of Fig. 8.24 uses just two straight lines. A horizontal one at gain 21 below 5 kHz and a descending one above 5 kHz. This is quite accurate well away from the break frequency of 5 kHz, but only an approximation at the break frequency itself. The gain does not suddenly start to drop when you go above 5 kHz, but the difference between the real graph and the two-line approximation is not very great when shown on log–log axes.

Active bass cut

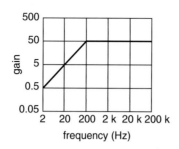

Fig 8.26 Gain–frequency graph for an active bass-cut filter.

This section shows you how to go about designing a filter with the gain–frequency graph shown in Fig. 8.26. The first step is to realize that an active filter will be needed. This is because the voltage gain is greater than one for high frequencies, requiring the use of an op-amp. Next, consider the equivalent circuits at frequencies above and below the break frequency of 200 Hz. These are shown in Fig. 8.27.

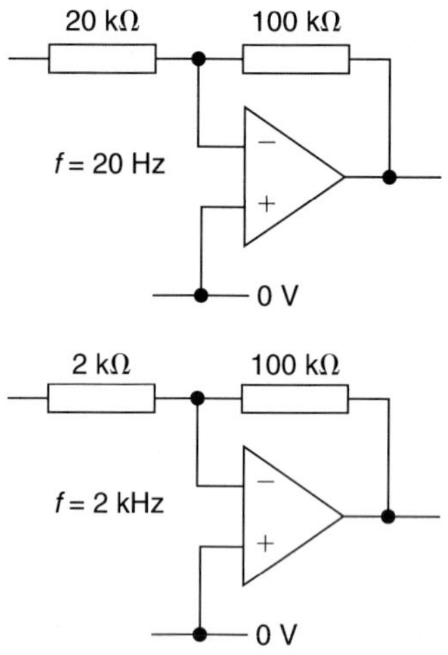

The gain–frequency graph is flat for frequencies above 200 Hz, so you can use this region to set values for the two resistors in the circuit. Let's set the input resistor to an arbitrary 2 kΩ. We can now calculate the value of the feedback resistor.

$R_f = ?$

$R_{in} = 2$ kΩ $G = \dfrac{R_f}{R_{in}}$,

$G = 50$

therefore $R_f = G \times R_{in} = 50 \times 2 \times 10^3 = 100 \times 10^3 = 100$ kΩ

This gives the equivalent circuit shown at the bottom of Fig. 8.27. Now you can decide where to place the frequency-dependent component, the capacitor. Consider the equivalent circuit at 20 Hz, well below the break frequency, as shown at the top of Fig. 8.27. At this frequency the gain needs to be reduced from 50 to 5 (see Fig. 8.26), so R_{in} needs to increase to 20 kΩ.

Fig 8.27 Equivalent circuits below and above the break frequency.

$R_{in} = ?$

$G = 5$ $G = \dfrac{R_f}{R_{in}}$,

$R_f = 100$ kΩ

therefore $R_{in} = \dfrac{R_f}{G} = \dfrac{100 \times 10^3}{5} = 2 \times 10^4$ Ω or 20 kΩ

Since the impedance of a capacitor increases with decreasing frequency, putting one in series with the input resistor does the trick as shown in Fig. 8.28. At 20 Hz, a capacitor with an impedance of 20 kΩ will dominate the value of R_{in}, giving a gain of about 5. The graph of Fig. 8.26 shows that the break frequency is 200 Hz. This can be used to calculate a value for the capacitor.

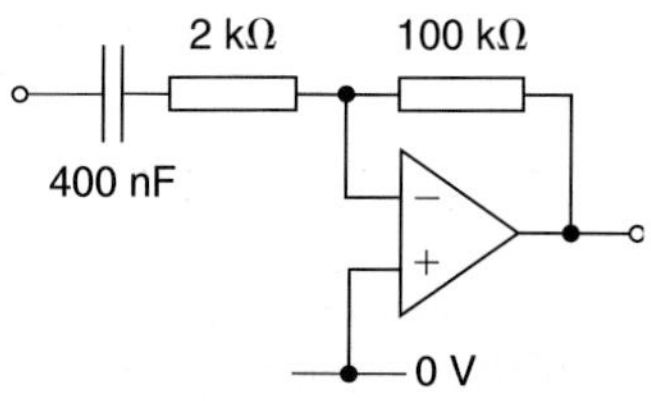

$C = ?$

$R = 2$ kΩ $f_0 = \dfrac{1}{2\pi RC}$,

$f_0 = 200$ Hz

therefore $C = \dfrac{1}{2\pi R f_0} = \dfrac{1}{2\pi \times 2 \times 10^3 \times 200} = 4.0 \times 10^{-7}$ F or 400 nF

Fig 8.28 Active bass-cut filter.

Amplifying audio

Bandpass filter

The circuit in Fig. 8.29 cuts out both high and low frequencies. It is a treble-cut filter and a bass-cut filter rolled into one circuit. But which components of the circuit determine the various characteristics of its gain–frequency graph?

The easiest point to start is the central region where the gain is flat. This means that the frequency-dependent components (the capacitors) have no effect on the gain for this range of frequencies, so they can be ignored. The flat gain can therefore be calculated from just the resistors.

$G = ?$

$R_{in} = 10 \text{ k}\Omega$

$R_f = 100 \text{ k}\Omega$

$$G = \frac{R_f}{R_{in}} = \frac{100 \times 10^3}{10 \times 10^3} = 10$$

Now consider the break frequency at 16 Hz. This is determined by the components which make up the input resistor of the amplifier. As the frequency decreases below 16 Hz, the impedance of the capacitor becomes much larger than the 10 kΩ resistance of the resistor, reducing the value of G.

$f_0 = ?$

$R = 10 \text{ k}\Omega$

$C = 1 \text{ }\mu\text{F}$

$$f_0 = \frac{1}{2\pi RC} = \frac{1}{2\pi \times 10 \times 10^3 \times 1 \times 10^{-6}} = 16 \text{ Hz}$$

The other break frequency at 16 kHz is the point at which the impedance of the 100 pF capacitor is 100 kΩ, the same as the resistor in parallel with it.

$f_0 = ?$

$R = 100 \text{ k}\Omega$

$C = 100 \text{ pF}$

$$f_0 = \frac{1}{2\pi RC} = \frac{1}{2\pi \times 100 \times 10^3 \times 100 \times 10^{-12}} = 1.6 \times 10^4 \text{ Hz}$$

At frequencies above 16 kHz, the impedance of the capacitor drops below 100 kΩ, reducing the size of the effective feedback resistor, resulting in a reduction of gain.

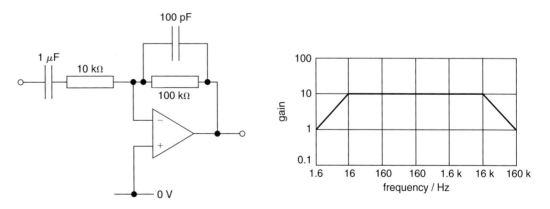

Fig 8.29 Gain–frequency graph for a bandpass filter.

Questions

8.1 Audio systems

1. A public address (PA) system processes the sound of a voice so that it can be heard over a wide area.

   (a) Draw a block diagram for the PA system. Use these blocks: loudspeaker, microphone, power amplifier, pre-amplifier, volume control, voltage amplifier.

   (b) Compare the input and output signals of each block.

2. The circuit of Fig. Q8.1 shows a microphone circuit and an inverting amplifier.

   (a) The output impedance of the microphone circuit is 6.8 kΩ. State the input impedance of the inverting amplifier.

   (b) Calculate the percentage loss of signal as it transfers from the microphone circuit to the amplifier. Assume a signal amplitude of 25 mV from the microphone.

   (c) Suggest how the system could be improved to avoid this signal loss.

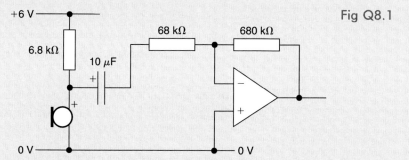

Fig Q8.1

3. Audio playback systems, such as CD players, are often equipped with tone controls.

   (a) Describe the function of a tone control.

   (b) With the help of a gain–frequency graph, suggest why bass boost may sometimes be necessary.

   (c) Use gain–frequency graphs to describe the effect of a treble boost filter on signals in the audio frequency range.

4. The block diagram in Fig. Q8.2 is part of an audio amplifier system.

   (a) The pre-amplifier has a voltage gain of +5 and an input impedance of 680 kΩ. Draw a suitable circuit based on an op-amp. Show all component values and justify them.

   (b) Draw a circuit for the volume control.

   (c) The voltage amplifier has a voltage gain of −40 and an input impedance of 33 kΩ. Draw a suitable circuit based on an op-amp. Show all component values and justify them.

   (d) Explain why the voltage gain of the whole system is between 0 and −200.

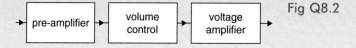

Fig Q8.2

Amplifying audio

5 The output of a JCB74 power amplifier is connected to an oscilloscope. The amplitude of the signal on the screen is 2.4 V. The amplitude decreases when a 16 Ω loudspeaker is also connected to the output of the JCB74.

    (a) The output impedance of the JCB74 is 4 Ω. Calculate the amplitude of the voltage across the loudspeaker.

    (b) Calculate the power of the loudspeaker.

8.2 Filters

1 This question is about the filter circuit in Fig. Q8.3.

    (a) Show that the break frequency of the filter is 200 Hz.

    (b) Explain why the gain of the filter is 1 at frequencies well below 200 Hz.

    (c) Draw a gain–frequency graph for the filter over the range 2 Hz to 20 kHz.

2 A passive bass-cut filter contains a 100 nF capacitor and has a break frequency of 60 Hz.

    (a) Describe how the impedance of a capacitor depends on the frequency of the signal across it.

    (b) Draw a circuit diagram of the filter. Show all component values and justify them.

    (c) Draw a gain–frequency graph for the filter over the range 6 Hz to 6 kHz. Use the impedance of the capacitor to explain the shape of the graph.

3 This question is about the active filter shown in Fig. Q8.4.

    (a) What are the advantages of using an active filter instead of a passive one?

    (b) Show that the break frequency of the filter is 1.5 kHz.

    (c) Explain why the gain is 14 for frequencies below 1.5 kHz.

    (d) Draw a gain–frequency graph for the filter over the range 15 Hz to 15 kHz.

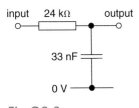

Fig Q8.3

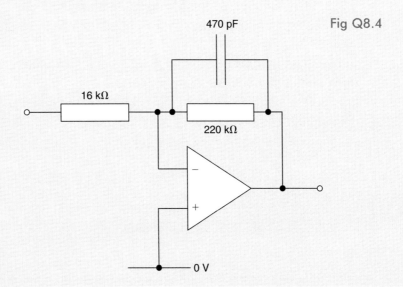

Fig Q8.4

4 Design a filter with the transfer characteristic in Fig. Q8.5. Show all component values and justify them.

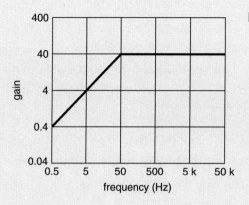

Fig Q8.5

5 A bandpass filter is required to behave as follows:
   - A constant gain of 30 for signals between 120 Hz and 12 kHz.
   - Signals below 120 Hz and above 12 kHz have gains of less than 30.

   (a) Draw a circuit diagram for the filter. Show all component values and justify them.

   (b) Draw a gain–frequency graph for the filter over the range 1.2 Hz to 120 kHz.

   (c) Use the impedance of the capacitors to explain the shape of the graph.

Learning summary

By the end of this chapter you should be able to:

- draw a block diagram for a complete audio system
- explain the function of the blocks in an audio system
- use the ideas of input and output impedance to improve signal transfer between blocks
- draw circuit diagrams for voltage and power amplifiers
- draw circuits for volume controls and filters
- analyse the performance of filter circuits and draw their transfer characteristics
- design filter circuits for use in audio systems

CHAPTER 9

Microcontrollers

9.1 Programmable systems

Both of the circuits shown in Fig. 9.1 have the same function. Both use a thermistor to monitor the temperature and turn on the heater if the temperature is below a predetermined level. The inputs and outputs are the same, but the processors are different. The left-hand processor is an op-amp, the right-hand one is a **programmable integrated circuit** or PIC.

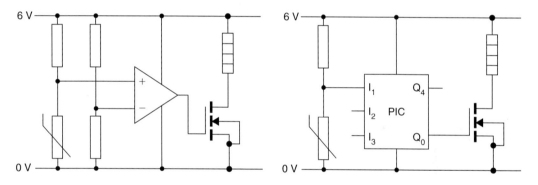

Fig 9.1 Both circuits have the same function.

The idea that the same function can be delivered by different integrated circuits should come as no surprise. After all, a wide variety of op-amps are available, each with different response times, supply requirements, output power, price, input current . . . – all determined by the arrangement of transistors and resistors on the chip. What makes the PIC different from all op-amps is that its behaviour is determined by a sequence of bytes stored in its memory. This **program** is shown as a **flowchart** in Fig. 9.2.

Fig 9.2 Program flowchart for the PIC of Fig. 9.1.

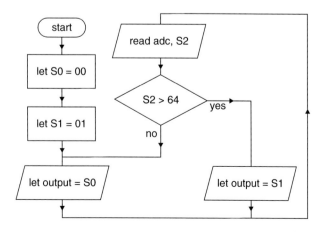

OCR Electronics for AS

Flowcharts

You will find out how to analyse and synthesize flowcharts later on in this chapter. For now, you only need to appreciate the following:

- A flowchart shows the sequence of operations required by the PIC.
- You can draw the flowchart with a computer running appropriate software.
- The computer can translate the flowchart into **machine code** (the required sequence of bytes).
- The sequence of bytes can be transferred from the computer to the PIC.

Once the bytes are stored safely in the memory of the PIC, it can be disconnected from its **host computer** and placed in the circuit. Each time the PIC is powered up it will run the program.

Hardware, software

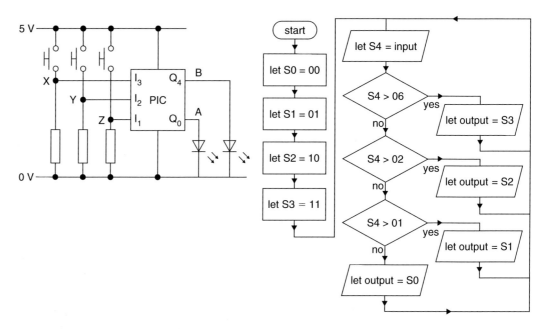

Fig 9.3 Using a PIC to make a priority encoder.

X	Y	Z	B	A
0	0	0	0	0
1	1/0	1/0	1	1
0	1	1/0	1	0
0	0	1	0	1

The circuit in Fig. 9.3 shows the effect of changing the program of a PIC. The **hardware** of the PIC (the thousands of logic gates on the integrated circuit) has not changed, only the **software** (the sequence of bytes stored in the memory). The different software makes the PIC into a logic system with this truth table.

The output codes, in binary, for the switch which has been pressed, with each switch having a different priority. So if X is high, the signals at Y and Z have no effect. Similarly, provided that X is low and Y is high, the state of Z is unimportant. Such systems are known as **priority encoders** and are widely used in complex systems, such as computers, to decide the order in which competing tasks must be performed.

Microcontrollers

PICs or gates?

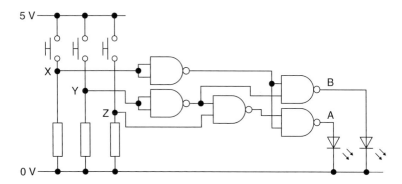

Fig 9.4 Using NAND gates to make a priority encoder.

The circuit of Fig. 9.4 uses five NAND gates to make a three-bit priority encoder which has exactly the same function as the one discussed on the previous page. As such, its behaviour is completely fixed by the arrangement of its hardware. It has to obey these expressions.

$$B = X + Y$$

$$A = X + \bar{Y}.Z$$

NAND gates come four to a chip, so implementing the circuit in Fig. 9.4 requires the use of two integrated circuits. This is about the half the cost of using a single small PIC to do the same job. However, there are other costs. A printed circuit board (p.c.b.) has to be designed, built and tested. Since the same PIC can be used to make a wide variety of logic systems, its p.c.b. layout only needs to be designed and tested once. This can save a lot of money. Here are some other reasons why you might use a PIC to make a logic system.

- The behaviour of the programmed PIC can be simulated by the host computer, allowing flaws in the program to be spotted before anything has been built.

- There is no need to simplify Boolean algebra expressions as part of the design process.

- PICs come in a variety of sizes, with up to eight inputs and eight outputs, so many NAND gate chips can often be replaced with a single PIC.

- Using a PIC allows you to adjust the behaviour of the system sometime in the future by simply changing its program.

You would probably have to replace the whole circuit board if you needed to adjust the behaviour of a logic system made from just NAND gate chips. However, in applications where speed of response is important, you should stick to NAND gates for logic systems. This is because a PIC takes time to step through the program, requiring about a microsecond to execute each instruction. This means that the response time of a PIC is often much slower than that of a NAND gate system.

More functions

Fig 9.5 The LED flashes on and off when the heater is on.

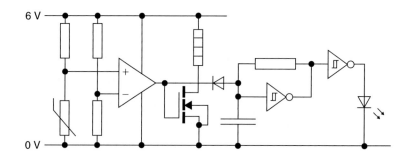

The use of a PIC often allows a system to take on extra tasks for very little extra cost. Consider the automatic heater circuit shown in Fig. 9.5. It is a variant of the circuit of Fig. 9.1, with the addition of a flashing LED to indicate that the heater is on. Extra hardware, in the form of a Schmitt trigger chip, a resistor, capacitor and LED are required for this. This inevitably adds to the cost of the system. Fig. 9.6 shows the addition of an LED to the spare output pin of the PIC allowing it to have the same behaviour as the circuit above, without the need for any extra chips, resistors or capacitors. The only change needed is some extra lines of code in the program.

Fig 9.6 Changing the program allows extra functions.

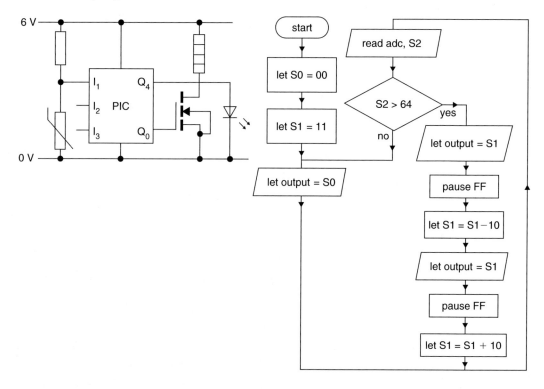

Economies of scale

Because of their versatility and ease of use, PICs are made in vast quantities. This makes them relatively cheap, despite the fact that they contain thousands of logic gates and are very expensive to design in the first place. PICs are widely used to control things (hence their alternative name of **microcontroller**), either as stand-alone systems (such as refrigerator thermostats, greetings cards or car engines) or embedded in much larger systems (such as games consoles, computers and telephone networks).

Microcontrollers

9.2 Hardware

This section will attempt to give you some idea about how the sub-systems inside a microcontroller interact to accomplish its marvellous flexibility and universality. However, do not expect the full story; fine details will be reserved for later, when you study microprocessors at A2. A top-down approach will be used, starting off with the whole system, then considering the function of each sub-system in turn.

The system

Fig. 9.7 shows the connections to a typical microcontroller integrated circuit. The pins fall into the following six different categories:

- Two power supply connections to the standard logic system values of +5 V and 0 V.

- A pair of serial input and output pins used to download the program in machine code from the host computer.

- A single reset terminal. This is usually active low; pull it low and then return it high to start runnning the program stored in memory.

- Eight digital input pins, labelled I_0 to I_7, allowing one byte of information to be transferred into the PIC.

- Eight digital output pins. labelled Q_0 to Q_7, allowing the PIC to latch a byte of information for inspection by other systems.

- The input to an analogue-to-digital converter (adc).

Some of these connections will be common to all PICs, regardless of their size. For example, all PICs need a power supply and terminals for downloading their programs. However, the number and type of input and output terminals varies from one PIC to another depending on its intended application. The rest of this chapter (and your AS exam) will assume a PIC with terminals as shown in Fig. 9.7.

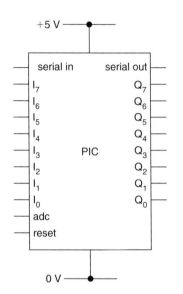

Fig 9.7 A typical microcontroller.

Registers

A microcontroller achieves its function by transferring bytes of information from one register to another. You will recall that a **register** is an assembly of D flip-flops whose clock terminals are connected in parallel as shown in Fig. 9.8. The word present at the four inputs (D_3 to D_0) is latched at the four outputs (Q_3 to Q_0) each time a rising edge arrives at CK. The timing of these rising edges are coordinated by the **clock**, an oscillator inside the PIC – typically running at 4 MHz.

Fig 9.8 D flip-flops arranged to make a four-bit register.

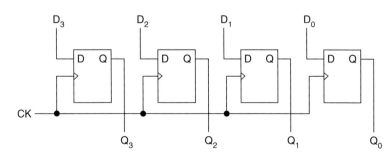

Ports

Digital information enters and leaves the PIC through **ports**. As you can see from Fig. 9.9, these can be the inputs and outputs of registers. So the signals at the eight input pins I_7 to I_0 are latched by the **input register** when required. Similarly, a byte from the **central processing unit** (or **CPU**) can be latched by the **output register**, where it can be accessed by other systems connected to Q_7 to Q_0.

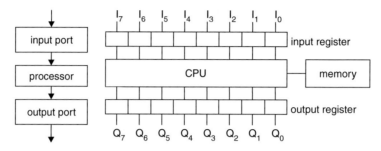

Fig 9.9 Simple block diagram for a microcontroller.

Memory

The sequence of operations to be carried out by the CPU is stored as a series of bytes in the **memory** of the microcontroller. You can think of the memory as a distinct sub-system, separate from the CPU as indicated by Fig. 9.9. Each byte is stored in a register at a unique **location** in the memory. The information about that location is coded as a binary word called the **address**. For example, the byte 11001110 might be stored at address 01010010.

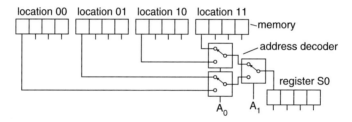

Fig 9.10 The address decoder selects which location provides the bit to be latched into the register.

The internal structure of a memory is complex. Fig. 9.10 shows a few of the connections in a memory with just four different locations. (A typical PIC will have 1024 locations!) Each location is a register which holds a four-bit word (a **nibble**, half of a byte). The one-of-four **address decoder** made from multiplexers uses the two-bit word A_1A_0 (the **memory address**) to direct the most significant bit (or msb) from just one of the four registers in the memory to the input of the register S0. That bit will latched by this register when a clock pulse arrives from elsewhere in the CPU. Of course, another three address decoders are required to latch the other three bits from the same location.

Expanding systems

A memory is a good example of how very large systems are designed in electronics. The trick is to design a small version of the system, such as the one of Fig. 9.10, using components which can be stacked together to make larger versions of themselves. So eight-bit registers can be made by stacking pairs of four-bit registers end to end, and one-of-eight address decoders can be made by adding another four multiplexers and an extra address input A_2. In this way, you can make such systems as large as you like with no extra design effort. The only problem is how to fit it all into the integrated circuit!

Microcontrollers

Processing words

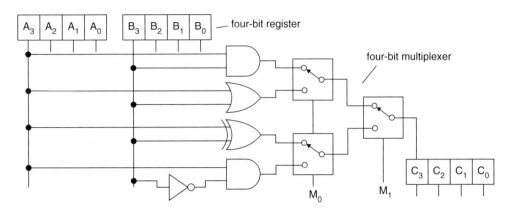

Fig 9.11 Pairs of bits from the A and B registers are processed and latched by the C register.

Fig. 9.11 attempts to show how a sequence of bytes stored in a microcontroller's memory are able to dictate its behaviour. Suppose that a two-bit word M_1M_0 is latched from the memory by a register which, for clarity, is not shown in the diagram. That word determines which set of logic gates provides a bit to be latched by the C register on the right. Each of the four sets has the same inputs, namely the msb of the A and B registers on the left, but performs a different function, as shown in the table.

memory M_1M_0	function
11	C = A AND B
10	C = A OR B
01	C = A EOR B
00	C = A AND NOT B

So a word stored in memory is able, through the use of a series of multiplexers and logic systems in parallel, to alter words as they shuffle from one register to another upon the arrival of rising edges from the system clock. As with memories on the previous page, the system shown in Fig. 9.11 can be easily expanded to accommodate eight-bit words from the memory, giving $2^8 = 256$ different operations, and eight-bit registers which hold bytes instead of nibbles.

Analogue inputs

Many PICs are endowed with at least one **analogue-to-digital converter**. This useful sub-system uses the voltage at the adc input to create an eight-bit word which contains information about that signal. A popular choice is to break the incoming signal into 10 mV chunks, with a separate word for each chunk as shown in this table.

input (mV)	byte
0–9	0000 0000
10–19	0000 0001
20–29	0000 0010
30–39	0000 0011
...	...
2530–2539	1111 1101
2540–2549	1111 1110
2550 and over	1111 1111

Two-bit conversion

Fig. 9.12 shows an easily expandable system which encodes an incoming analogue signal V_{in} as a two-bit word. It functions as follows:

1. A resistor ladder generates reference signals at 1 V, 2 V and 3 V.
2. A triplet of op-amps compares the signal V_{in} at the adc input with these reference signals.
3. The outputs X, Y and Z of the op-amps go high or low according to the value of V_{in}, as shown in this table.

V_{in} (V)	X	Y	Z
0.00–0.99	1	1	1
1.00–1.99	0	1	1
2.00–2.99	0	0	1
3.00 and over	0	0	0

4. The EOR gates process X, Y and Z to produce signals L_3 to L_0, each of which goes high in a unique range of values of V_{in}.

V_{in} (V)	X	Y	Z	L_0	L_1	L_2	L_3
0.00–0.99	1	1	1	1	0	0	0
1.00–1.99	0	1	1	0	1	0	0
2.00–2.99	0	0	1	0	0	1	0
3.00 and over	0	0	0	0	0	0	1

5. Finally, a pair of OR gates combine the signals at L_3 to L_0 to generate the required binary code BA for the output.

V_{in} (V)	X	Y	Z	L_0	L_1	L_2	L_3	B	A
0.00–0.99	1	1	1	1	0	0	0	0	0
1.00–1.99	0	1	1	0	1	0	0	0	1
2.00–2.99	0	0	1	0	0	1	0	1	0
3.00 and over	0	0	0	0	0	0	1	1	1

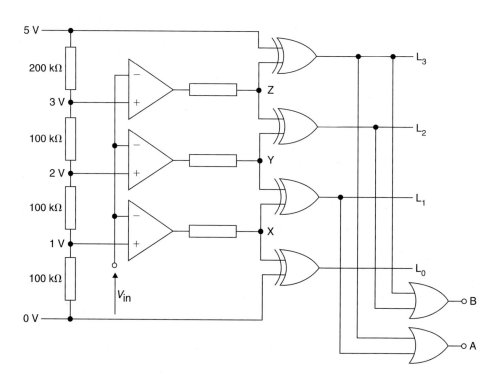

Fig 9.12 An expandable two-bit analogue-to-digital converter.

Microcontrollers

9.3 Software

Writing good software for microcontrollers is an art which is acquired by practice. You only get to be good at it by doing it lots of times. This section will show you how a limited set of flowchart symbols can be used to write a wide variety of programs, making PICs perform many useful tasks. There will be no attempt to go beyond these basics; you will find out the full story when you study microprocessor systems at A2.

Registers

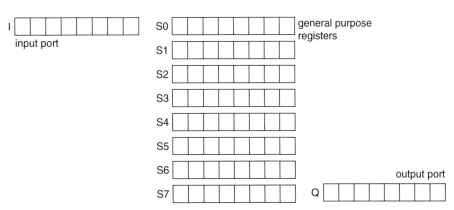

Fig 9.13 Programs control ten registers in the PIC.

This section is going to assume that the PIC has the following registers, shown in Fig. 9.13:

- An eight-bit input port (I).
- An eight-bit output port (Q).
- Eight general purpose eight-bit registers (S0 to S7).

Programs will be presented as flowcharts. These will represent, in pictorial form, the sequence of changes made to the binary words stored in these registers inside the PIC. Keeping track of all eight bits in a binary word is difficult, so **hexadecimal coding** will be used throughout. The eight bits in a byte are split into a pair of nibbles, each of which can then be represented by a single character, as shown in the table. So the byte 01111100 is split into the nibbles 0111 and 1100. As you can see from the table, the nibble 0111 is coded as 7 in hexadecimal (or **hex**) and 1100 is coded as C. So 01111100 in binary is 7C in hex.

decimal	binary	hex
0	0000	0
1	0001	1
2	0010	2
3	0011	3
4	0100	4
5	0101	5
6	0110	6
7	0111	7
8	1000	8
9	1001	9
10	1010	A
11	1011	B
12	1100	C
13	1101	D
14	1110	E
15	1111	F

OCR Electronics for AS

PICs count

Fig. 9.14 shows a very straightforward microcontroller circuit. A set of eight LEDs display the state of the PIC's output port. The program is shown as a flowchart to the left of the circuit. The picture shows a number of different shaped boxes connected by arrowed lines. Each box represents a different instruction, with its shape and contents telling you what effect that instruction has on the registers. The arrows tell you the order in which the instructions are to be obeyed.

You can **analyse** the effect of the flowchart on the PIC as follows:

- The **start** box indicates the first instruction to be obeyed after the PIC has been reset.

- The first instruction is in a square **process** box. It resets the contents of the register S0 to zero.

- The parallelogram **output** box copies the contents of S0 to the output port Q.

- Another process box makes the PIC wait for 200 ms before moving to the next instruction. The hex word C8 is 1100 1000 in binary and $12 \times 16 + 8 = 200$ in decimal.

- The next instruction adds one to the contents of the S0 register.

- This new byte is copied to the output port . . . .

The overall effect of the program is to make the system display, in binary, all of the numbers from 0 to 255, in ascending order, with a pause of 200 ms between changes. It does this over and over again until the power supply is disconnected.

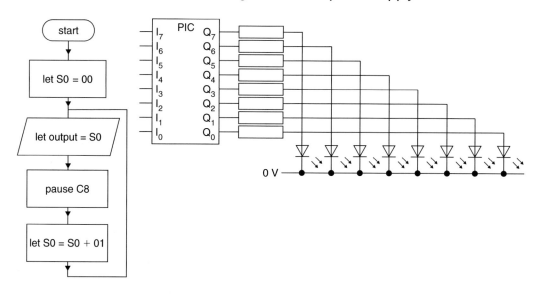

Fig 9.14 The LEDs continuously count up in binary.

Microcontrollers

Decisions

What makes systems based on PICs special is their ability to make decisions. This is illustrated by the system shown in Fig. 9.15. The program makes the display continuously show the numbers from 0 to 9, one after the other, changing four times a second.

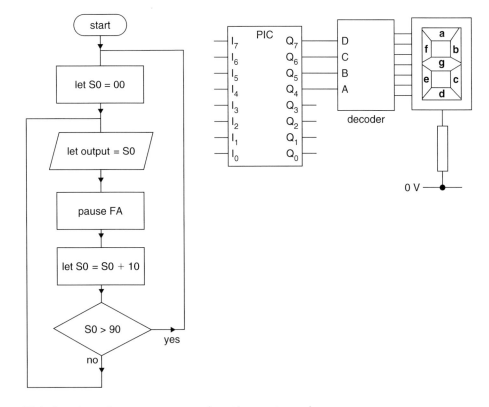

Fig 9.15 The decision box makes the display reset to zero after nine.

This is what the program makes the system do:

1. The S0 register is reset and then copied to the output port. The decoder feeds an appropriate seven-bit word abcdefg to the seven segment display, showing the number zero.

2. The system waits for 250 ms. (FA is $15 \times 16 + 10 = 250$ in decimal).

3. The S0 register is increased by 10 (in hex).

4. The lozenge symbol shows that program then **branches**. If S0 is greater than 90 (in hex) then the system starts the whole program again, otherwise the new byte in S0 is copied to the output port, resulting in a display of the number one.

Notice how the use of a decoder only requires the use of four bits of the output port, rather than all eight.

OCR Electronics for AS

Inputs

Fig 9.16 The system counts how often the switch is briefly pressed.

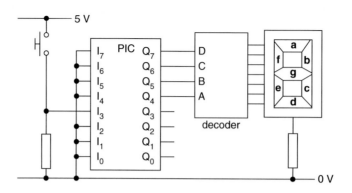

The PIC system shown in Fig. 9.16 uses a switch to accept information from the outside world. Bit I_3 is pulled low by a resistor unless the switch is pressed. All seven other bits have been pulled low by direct connection to the 0 V supply rail, preventing them from floating and producing erratic signals. The flowchart of Fig. 9.17 shows the program stored in the PIC. This is what it makes the system do:

1 Show the number zero on the seven segment display.

2 Copy the input port to register S1.

3 Continually repeat the last step if S1 is not 08. The input port is only 08 (in hex) when the switch is pressed, making the word at the input port 0000 1000 (in binary). So this instruction makes the system wait until the switch is pressed.

4 Increase the most significant nibble of S0 by one.

5 Wait for 250 ms, allowing time for the switch to be released.

6 Copy S0 to the output port, displaying the next number up, unless that number is ten. If S0 is A0, the program goes back to its first instruction.

7 Wait until the switch is pressed again . . . .

The fourth and fifth instructions make a **test-and-skip loop**. This program technique is widely used to make PIC systems wait until one specific byte is present at the input port.

Fig 9.17 Program for the system in Fig. 9.16.

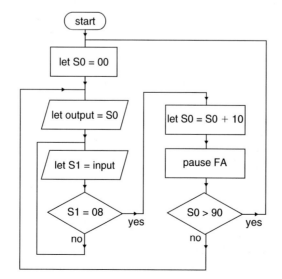

Microcontrollers

Programmable monostable

A monostable produces a single low pulse of fixed length when it is triggered by a falling edge. The system of Fig. 9.18 is required to behave like a monostable, with two differences:

- The output pulse is triggered by a rising edge.
- The pulse duration is determined by the nibble DCBA at the input port, as shown in this table.

DCBA	time (ms)
0000	0
0001	1
0010	2
...	...
1110	14
1111	15

Fig 9.18 A programmable monostable.

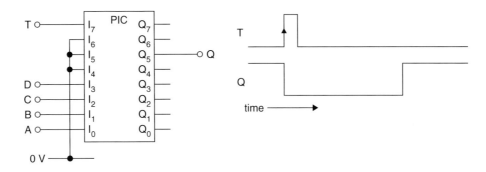

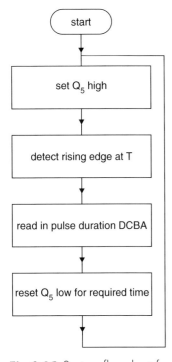

Fig 9.19 System flowchart for the programmable monostable.

The first stage in writing the program is to draw a **system flowchart** like the one shown in Fig. 9.19. This shows the sequence of operations that the system must perform in order to accomplish its function. In the case of a programmable monostable, the sequence falls into four distinct **chunks**. Each chunk (or **routine**) can then be written (and tested if necessary) as a separate flowchart, before linking them together to make the whole program. This **modular** approach to PIC programming is widely adopted because it makes the whole task easier and less prone to error.

Initialization routine

The flowchart chunk shown Fig. 9.20 resets Q_5 high by loading register S0 with 0010 0000 (in binary) and then copying this to the output port. Notice the two **links** to other flowcharts, shown as circles with labels inside them. The direction of the arrow next to each circle tell you which way the links go. So link **a** takes the program control **into** another flowchart with a link also labelled **a** (Fig. 9.21). Similarly, link **d** takes the control **from** another flowchart with a link labelled **d** (Fig. 9.23).

Fig 9.20 Setting Q_5 high.

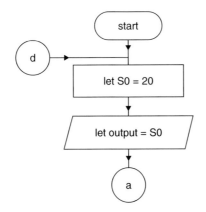

OCR Electronics for AS

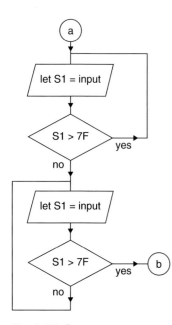

Detecting rising edges

The flowchart chunk shown in Fig. 9.21 consists of a pair of test-and-skip loops. The first loop makes the system wait until I_7 is low. When this is so, the byte copied to S1 cannot be greater than 0111 1111 (in binary) or 7F (in hex). The next loop makes the system wait until the instant that I_7 goes high. When this happens, the byte copied to S1 is greater than 0111 1111 (in binary) or 7F (in hex), so program control flows to link **b**.

Pulse duration

The pulse duration DCBA is fed into the system as the least significant nibble of the input port. This needs to be separated from the bit at I_7. Since this bit is going to be high when system is triggered, subtracting 1000 0000 from the byte at the input port should leave 0000 DCBA. This processing is performed by the chunk shown in Fig. 9.22, leaving the result in register S2.

Going low

The final flowchart chunk is shown in Fig. 9.23. The first step tests S2 to see if it is 00. If not, S0 is reset to 00 and copied to the output port, lowering Q_5 to 0. After a delay of 1 ms, the byte in S2 is reduced by one, before returning to the first step of the chunk. Eventually, after a number of cycles round the loop, S2 will be 00 and control will pass through link **d** to the start of the whole program.

Fig 9.21 Detecting a rising edge at I_7.

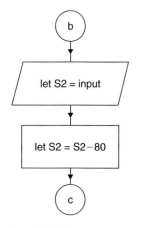

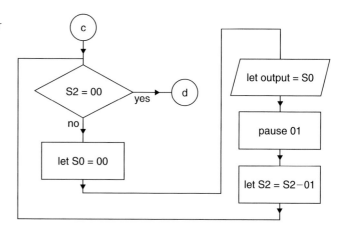

Fig 9.23 Pulling Q_5 low for required time.

Fig 9.22 Reading in the pulse duration.

Truth tables

PICs are often a good way of implementing complex truth tables, provided that you are able to accept their relatively slow response time. The system flowchart of Fig. 9.24 shows the three steps required by the software to implement a truth table:

1 Load the output words of the truth table into registers.

2 Determine which input word of the truth table is present at the input port.

3 Copy the appropriate register to the output port.

The last two steps are then repeated until the system is reset or the power supply fails.

Microcontrollers

Fig 9.24 System flowchart for a PIC programmed as a two-bit to seven-segment decoder.

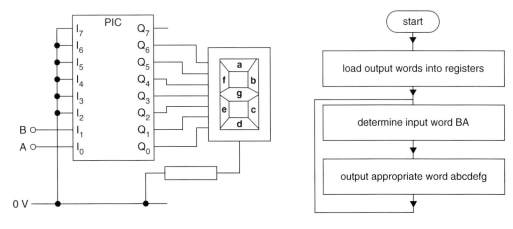

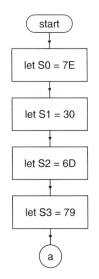

Fig 9.25 Loading the output of the truth table into registers.

Output words

The system of Fig. 9.24 has to display, in decimal, the binary equivalent of the two-bit word BA present at the input port. A truth table, showing the required output word for each input word makes a good starting point.

input BA	display	output abcdefg
00	zero	01111110 (7E)
01	one	00110000 (30)
10	two	01101101 (6D)
11	three	01111001 (79)

Notice how each bit of the output word is chosen to make the seven-segment display show the appropriate number. The first chunk of the program, shown in Fig. 9.25, loads the four output bytes into the registers S0, S1, S2 and S3.

Decision ladder

Fig. 9.26 shows a **decision ladder** which can detect which of the four possible words BA is present at the input port. Notice that when the first three possibilities have been rejected, the fourth is assumed. The ladder passes program control to one of four chunks through the links **b**, **c**, **d** and **e**. Each of these chunks copies a different register to the output port before passing program control back to the start of the decision ladder (Fig. 9.27).

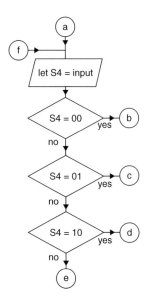

Fig 9.26 Determining the input word with a decision ladder.

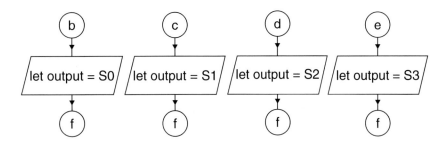

Fig 9.27 Each flowchart chunk copies a different word to the output port.

Analogue inputs

Although a PIC is a digital system, it can also deal with analogue signals through its analogue-to-digital converter (adc) input. For example, the PIC system shown in Fig. 9.28 has been programmed to act as a light meter, translating the varying light conditions of the LDR to a three-bit word displayed through a trio of LEDs.

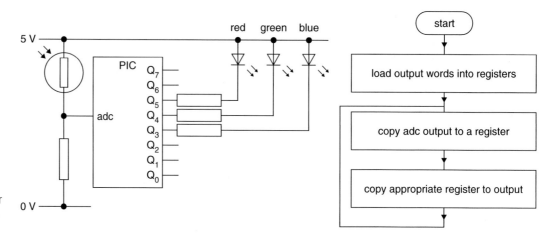

Fig 9.28 System flowchart for a PIC programmed as a light meter.

The specification of the system is given in this table.

LDR illumination	input voltage (V)	output display
very bright	1.81–2.55	only red
bright	1.51–1.80	red and green
just right	0.91–1.50	flashing green
dim	0.61–0.90	green and blue
very dim	0.00–0.60	only blue

The system is required to act as a light meter, possibly for a camera where correct illumination of the subject is vital for good pictures. The ideal illumination of the LDR places a voltage between 0.91 V and 1.50 V at the adc input of the PIC, resulting in a flashing green LED. Increasing the illumination a bit stops the LED flashing and brings on the red one as well. Further increasing the illumination turns off the green LED altogether. Similarly, the blue LED indicates that the light getting to the LDR is very dim, and the green and blue LEDs together indicate dim light.

Output words

The adc eight-bit output word hgfedcba is determined by the input voltage V as follows:

$$V \times 100 = hgfedcba \text{ expressed as a decimal between 255 and 0}$$

The table shows the required output words to be stored in registers for each range of output words from the adc. Notice that an output has to be pulled low to make an LED glow.

Microcontrollers

input voltage (V)	adc output	output word
1.81–2.55	B5–FF	00011000 (18)
1.51–1.80	97–B4	00001000 (08)
0.91–1.50	5B–96	00101000 (28)
0.61–0.90	3D–5A	00100000 (20)
0.00–0.60	00–3C	00110000 (30)

Loading registers

The flowchart in Fig. 9.29 shows the five different output words loaded into the registers S0 to S4. Each word is destined to be copied to the output port by another flowchart chunk. The word loaded into S2 corresponds to the voltage range where the green LED flashes. This is a problem which will be dealt with below, but for the moment S2 is loaded with a word which will turn only the green LED on.

Making decisions

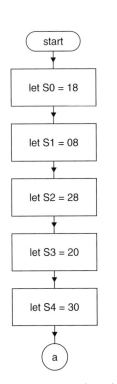

Fig 9.29 Loading the output words into registers.

Fig 9.30 Decision ladder to determine the input voltage.

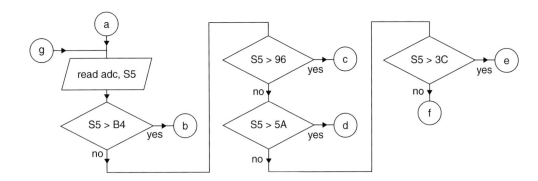

The flowchart chunk of Fig. 9.30 instructs the adc to copy its output into register S5, followed by a four-stage decision ladder. Each decision box compares the byte in S5 with the adc byte at the bottom of each voltage range, starting with the highest. So very bright light on the LED causes program control to pass through link **b**, whereas very dim light causes program control to pass through link **f**.

Steady outputs

Passing program control through links **b**, **c**, **e** or **f** results in a steady glow from one or two LEDs. The four flowchart chunks which do this are shown in Fig. 9.31. Program control flows out of each chunk through link **g** to the start of the decision ladder, so that the system continually updates the LED display to keep track of changing light levels on the LED.

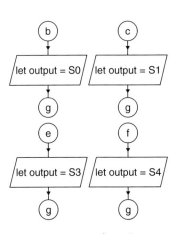

Fig 9.31 Output flowchart chunks for the output words which are steady.

OCR Electronics for AS

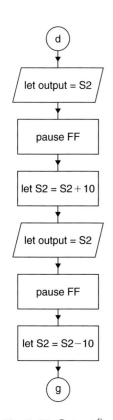

Fig 9.32 Output flowchart chunk which toggles the green LED.

Toggling bits

The specification of the system requires that the green LED flashes on and off if the byte in S5 is between 96 and 5A. This condition results in program control leaving the decision ladder through link **d** to the flowchart chunk shown in Fig. 9.32.

This chunk performs the following operations:

1 Output the word 28 (0010 1000), turning on only the green LED.

2 Wait for 255 ms.

3 Add 10 to the contents of S2 (0001 0000 + 0010 1000 = 0011 1000).

4 Output the word 38 (0011 1000) turning all three LEDs off.

5 Wait for another 255 ms.

6 Subtract 10 from S2 to restore its word to 28 (0010 1000).

7 Return to the start of the decision ladder.

The chunk **toggles** one bit of the output port (Q_4) without altering the state of the other bits. The green LED comes on and off again. If the voltage at the input of the adc remains in the target range of 0.91 V to 1.50 V, repeated passes through this flowchart chunk results in the green LED flashing on and off at about 2 Hz.

The end

There is one last flowchart symbol for you to meet. The **stop** box is rarely used in real flowcharts because its function is to halt the program flow. A PIC which has stopped running a program is hardly ever useful! However, the system in Fig. 9.33 shows the **stop** command in action. Immediately after a reset, the PIC energizes the lock and then waits until T goes high. This gives someone time to place the correct entry code GFEDCBA = 1101011 at the pins of the input port. When ready, they pulse T high. This makes the program control pass to a decision box. If the entry code is correct, program flow passes to link **a**, and another part of the program resets Q_1 and opens the lock. However, an incorrect code results in program flow being halted, so that the lock can never be opened . . . until the system is reset.

Fig 9.33 Partial flowchart for a PIC acting as a security system.

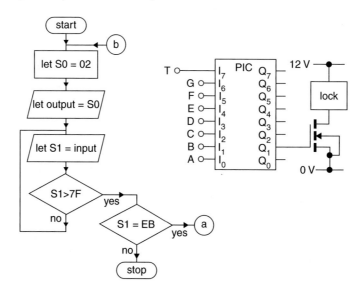

Microcontrollers

Questions

9.1 Programmable systems

1 A logic system can be implemented with NAND gates or with a microcontroller.

   (a) A microcontroller is a **programmable system**. Explain what this means.

   (b) Describe the steps required to implement a logic system with a microcontroller.

   (c) Explain one advantage of building a logic system out of NAND gates instead of a microcontroller.

2 The software loaded into the hardware of a microcontroller determines its behaviour.

   (a) Explain the meaning of the terms **software** and **hardware**.

   (b) Explain **three** advantages of implementing a digital system with a microcontroller instead of logic gates.

9.2 Hardware

1 A microcontroller contains a number of registers.

   (a) What is the function of a register?

   (b) Show how a four-bit register can be implemented with D flip-flops. Label the inputs and outputs.

   (c) Describe how to load a four-bit word into the register.

2 The PDB42 microcontroller has the following five sets of terminals:

   - A reset pin.
   - Serial input and output pins.
   - A two-bit input port.
   - A four-bit output port.
   - Two adc inputs.

   Describe the function of each of these sets of terminals.

3 The memory of a MRB21 microcontroller stores eight-bit words, with addresses ranging from 000000 to 111111.

   (a) Explain the meaning of the term **address**.

   (b) How many bits can be stored in the memory?

   (c) What is the function of the memory of a microcontroller?

OCR Electronics for AS

input (V)	output
0.00	00
0.01	01
0.02	02
...	...
2.54	FE
2.55	FF

Fig Q9.1

9.3 Software

Assume that the output of the adc is related to the voltage at its input as shown in this table.

1 Fig. Q9.1 shows the software loaded into a microcontroller.

   (a) What is the effect of the instruction **let S4 = 80**?

   (b) What is the effect of the instruction **read adc, S0**?

   (c) The decision box refers to the byte 1F. What is this byte in binary?

   (d) What is the effect of the instruction **let output = S7**?

   (e) Describe, in detail, the behaviour of the microcontroller when the program is run.

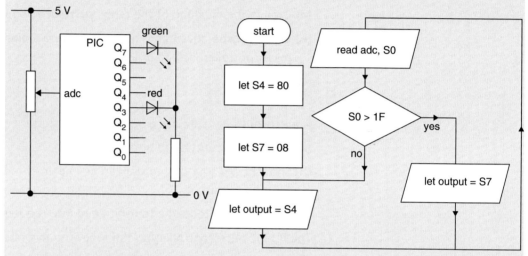

2 The microcontroller shown in Fig. Q9.2 is programmed so that the number of glowing LEDs is the same as the number of switches being pressed. An incomplete flowchart of the program is shown in Fig. Q9.3.

   (a) Write out a truth table to show the byte (in binary and hex) at the input port for each combination (pressed or released) of the switches.

   (b) Describe the effect on the system of the flowchart shown in Fig. Q9.3.

   (c) Complete the flowchart. Explain the function of each addition to the flowchart.

Fig Q9.2

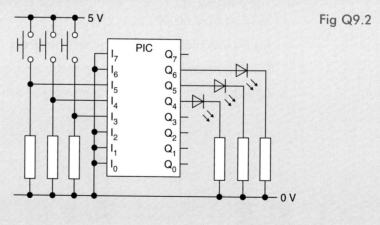

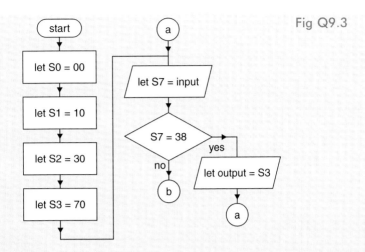

Fig Q9.3

3. This question is about the PIC shown in Fig. Q9.4. Whenever a particular four bit word DCBA is placed at the input port, P goes low for 42 ms before going high again.

   (a) Write a flowchart for the first part of the program which sets P high, then passes control to link **a**.

   (b) Describe the effect on the system of the flowchart between links **a** and **b**.

   (c) Write a flowchart for the last part of the program.

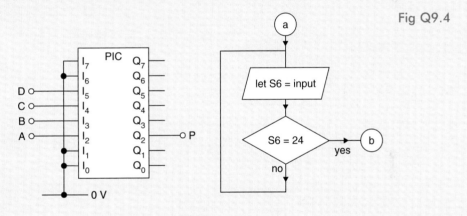

Fig Q9.4

OCR Electronics for AS

4 The PIC shown in Fig. Q9.5 is to behave as follows:
- Reset the output.
- Wait until only I_6 is high.
- Oscillate one pin of the output port through 60 cycles.
- Return to the start.

(a) Explain what effect the flowchart shown in Fig. Q9.5 has on the system.

(b) Which pin of the output port oscillates? At what frequency does it oscillate?

(c) Write a flowchart which makes the system execute the flowchart between links **a** and **b** sixty times before passing control to link **c**.

(d) Write a flowchart for the rest of the program.

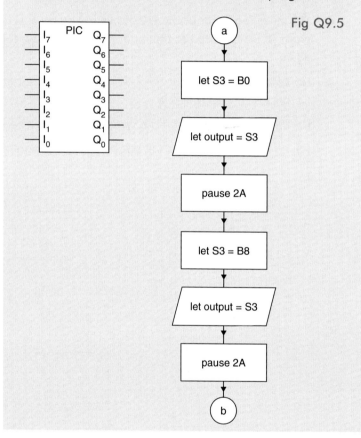

Fig Q9.5

Learning summary

By the end of this chapter you should be able to:
- describe a microcontroller system
- explain the difference between hardware and software
- state the advantages and disadvantages of programmable systems
- analyse and design simple microcontroller program flowcharts
- use hexadecimal code to represent bytes

Appendices

Formulae

Formula	Description
$R = \dfrac{V}{I}$	R is resistance in ohms (Ω) V is voltage drop in volts (V) I is current in amps (A)
$P = IV$	P is the power in watts (W) I is the current in amps (A) V is the voltage drop in volts (V)
$\tau = RC$	τ is the time constant in seconds (s) R is the resistance in ohms (Ω) C is the capacitance in farads (F)
$T = 0.7RC$	T is the pulse duration of a monostable in seconds (s) R is the resistance in ohms (Ω) C is the capacitance in farads (F)
$T = 0.5RC$	T is the period of a relaxation oscillator in seconds (s) R is the resistance in ohms (Ω) C is the capacitance in farads (F)
$f = \dfrac{1}{T}$	f is the frequency of a signal in hertz (Hz) T is the period of the signal in seconds (s)
$G = \dfrac{V_{out}}{V_{in}}$	G is the voltage gain of an amplifier V_{in} is the voltage of the input signal in volts (V) V_{out} is the voltage of the output signal in volts (V)
$V_{out} = A(V_+ - V_-)$	V_{out} is the output voltage of an op-amp in volts (V) A is the open-loop gain of the op-amp V_+ is the voltage at the non-inverting input in volts (V) V_- is the voltage at the inverting input in volts (V)
$G = 1 + \dfrac{R_f}{R_d}$	G is the voltage gain of a non-inverting amplifier R_f is the resistance of the feedback resistor in ohms (Ω) R_d is the resistance of the pulldown resistor in ohms (Ω)
$G = -\dfrac{R_f}{R_{in}}$	G is the voltage gain of an inverting amplifier R_f is the resistance of the feedback resistor in ohms (Ω) R_{in} is the resistance of the input resistor in ohms (Ω)
$-\dfrac{V_{out}}{R_f} = \dfrac{V_1}{R_1} + \dfrac{V_2}{R_2} \ldots$	V_{out} is the voltage at the output of a summing amplifier in volts (V) R_f is the resistance of the feedback resistor in ohms (Ω) V_n is the voltage at an input in volts (V) R_n is the resistance of an input resistor in ohms (Ω)
$f_0 = \dfrac{1}{2\pi RC}$	f_0 is the break frequency of a filter in hertz (Hz) R is the resistance of the resistor in ohms (Ω) C is the capacitance of the capacitor in farads (F)

OCR Electronics for AS

Circuit symbols

Fig A.1 Circuit symbols.

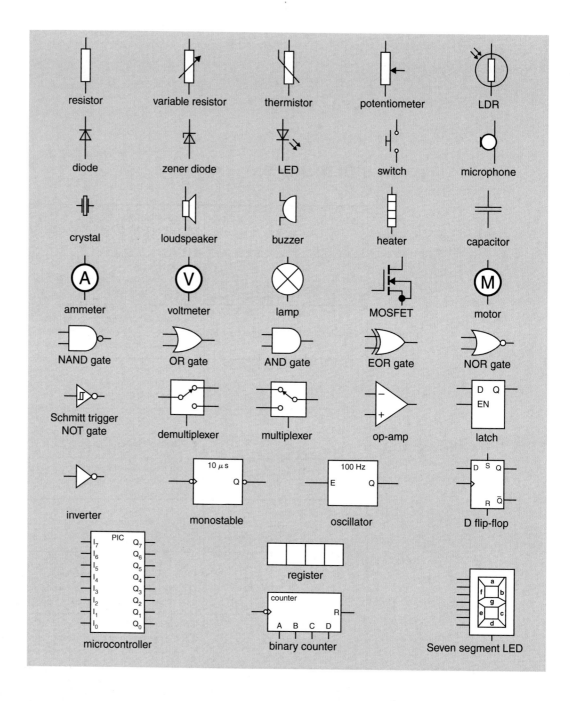

Appendices

Prefixes

prefix	multiplier
G	$\times 10^9$
M	$\times 10^6$
k	$\times 10^3$
m	$\times 10^{-3}$
μ	$\times 10^{-6}$
n	$\times 10^{-9}$
p	$\times 10^{-12}$

Boolean algebra

$$A.\overline{A} = 0$$

$$A + \overline{A} = 1$$

$$A.(B + C) = A.B + A.C$$

$$\overline{A.B} = \overline{A} + \overline{B}$$

$$\overline{A + B} = \overline{A}.\overline{B}$$

$$A + A.B = A$$

$$A.B + \overline{A}.C = A.B + \overline{A}.C + B.C$$

Flowchart symbols

There are eight general purpose registers Sn, where n = 0 to 7. The byte b is written in hexadecimal code.

Fig A.2 Flowchart symbols.

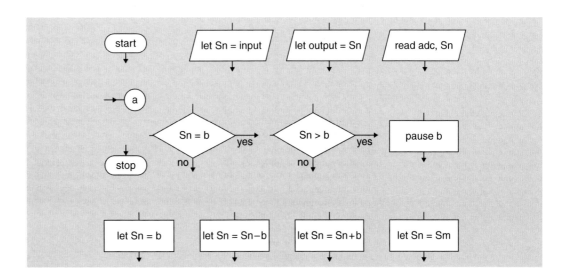

Index

address decoders 115, 142, 147, 151
algebra *see* Boolean algebra
alternating current (a.c.)
    amplifiers 83, 85, 93
    audio systems 120, 121, 123, 125, 127–8
    blocking 128
    microphones 83
ammeters 2, 160
amperes (amps/A) 3
amplifiers 82–99
    audio systems 121–5, 130
    coupling capacitors 83, 85
    current 83, 85, 87–8, 90–1, 93–4
    design 90, 92
    digital-to-analogue 95
    distortion 85
    ideal 84
    input resistors 91–2, 96
    inverting 91–2, 121, 123, 125, 130, 159
    known gain 89–92, 98
    linear region 84
    microphones 82–3
    mixers 96
    near-ideal 86
    non-inverting 89–90, 122–3, 159
    op-amps 82, 86–92, 121, 125–6, 130, 132
    peak current 87–8
    peak values 83, 84, 86, 90
    polarity 95
    power 120, 125
    resistors 83, 85, 87, 89–94, 96
    saturation levels 84, 85, 87, 89, 94, 96
    sine waves 85, 86
    summing 93–6, 99, 124, 159
    summing formula 94, 95, 159
    supply rails 91
    transfer characteristics 82, 84–5, 87–8, 92
    triangle waves 85
    virtual earth 91, 94
    voltage followers 86–8, 97–8, 125, 130
    voltage gain 84–92, 121–3, 125, 130–3, 159
    voltage gain formulas 91, 159
    voltage output formulas 94, 95, 159
amplifying audio 120–36
    audio systems 120–6, 134–5
    filters 124, 127–33, 135–6
analogue signals 19
    digital signals from 19–35
    hardware inputs 143, 144
    software inputs 152–4
analogue-to-digital converters 141, 143–4, 152–4
    see also digital-to-analogue converters
AND gates 8

binary counters 102, 103
    Boolean algebra 51, 54
    circuit symbol 160
    delaying signals 29
    flip-flops 76
    logic system design 54, 55
    truth tables 49, 50, 52, 54, 60
    using NAND gates for 60
AND rules 52, 53
angle sensors 22
anodes 25
audio systems 120–36
    amplifying audio 120–36
    bass boost 124
    block diagrams 120–1, 124
    equivalent circuits 121
    filters 124, 127–33, 135–6
    flat response 123, 124
    good matching 122
    howling 123
    impedance 121–2, 126–9, 131–3
    impedance matching 121–2
    input impedance 121, 122
    instability 123
    loudspeakers 120, 123–6
    microphones 120, 121, 122
    more power 125
    oscillation 123
    output impedance 121, 122, 126
    poor matching 122
    power amplifiers 120, 125
    power transfer 126
    pre-amplifiers 120–4
    short-circuit protection 126
    signal loss 122
    signal source 120, 121, 123
    stability 123
    supply rails 125
    switched control 124
    time constants 128
    tone control 120, 123–4
    treble boost 124
    voltage amplifiers 120, 124
    voltage drivers 122
    volume control 120, 124, 130

barchart sensors 26–7
bass boost 124
BCD (binary-coded decimal) 100
bias voltage 7
binary counters 100–4, 116
    cascading counters 104
    circuit symbols 160

Index

clocks 106–9
   continuous sequencers 110
   and decimal counters 103, 104
   decimal displays 103
   decoders 103
   four-bit counters 101, 113
   maximum count formula 102
   one-bit counters 100–1, 107, 109, 111
   one-shot sequencers 112–13
   one-to-five counters 102–3
   pulse tables 100–3
   reset 102, 107, 109–10, 112–13
   square waves 101, 108, 109
   three-bit counters 102
   timing diagrams 101, 103
   two-bit counters 101
binary words 8, 50, 78
   binary counters 100, 102–3
   eight-bit 143
   four-bit 100, 141, 142
   and programmable systems 145–7, 149–51, 154
   storage 72
   two-bit 143
binary-coded decimal (BCD) 100
bistables 67–71, 79
   active-high inputs 69, 73, 74
   active-low inputs 71
   algebraic analysis 70, 71
   feedback loops 67, 68
   forbidden states 70
   and latches 73
   NAND gate bistables 71
   NOR gate bistables 69–70, 73
   NOT gate bistables 67, 68
   pulsed high inputs 69
   set and reset 68, 72–4
   timing diagrams 68, 69, 70–1
   two stable states 67
bits
   binary counters 100–2
   storage 72
   toggling 154
block diagrams 12–14
   analysis 13
   audio systems 120–1, 124
   EOR gates 12–13
   information flow 12
   input blocks 12, 13
   light sensors 23
   NOR gates 14
   output block 12, 13
   processor block 12
   specifications 14
   synthesis 14
   truth tables 13–14
Boolean algebra 51, 139, 160
   bistables 70, 71
   brackets 51, 55, 57–8, 62, 70
   continuous sequencers 111
   designing logic systems from truth tables 54–8
   only NAND gates 59–63
   simplifying 51, 56
   sum of two terms 51
   table to algebra 53
brackets 51, 53, 55, 57–8, 62, 70, 88
break frequency 128, 129, 130–3, 159

breakdown voltage 26–7
butes 150
buzzers 160
bytes 138, 142, 143, 145

capacitors 29–32
   audio systems 123, 124, 127–9, 131–3
   bottom plates 29
   charging 29–30, 38, 41, 44, 128
   circuit symbols 160
   coupling capacitors 83, 85, 124, 127
   discharging 29, 31–2, 38–41, 44, 77, 128
   flip-flops 76, 77
   impedance 127–9, 131–3
   memory 31
   one-shot sequencers 113
   oscillators 44
   programmable integrated circuits 140
   spike-to-pulse conversion 40
   spikes 38–40
   switches and 31
   top plates 29–32
cathodes 25
central processing unit (CPU) 142
charge 2, 4
   changing 29
   diodes 25
   forward biased 25
   NOT gates 6
chunks (routines) 149–51, 153–4
circuit design 14
circuit diagrams 12
circuit elements 14
circuit symbols 12, 42, 53, 75, 160
clocks 105–9, 117
   counting hours 107
   counting seconds 106
   crystal oscillators 105–6, 111
   displays 105, 106
   frequency division 108–9
   microcontrollers 141, 143
   stopwatches 107
   sub-systems 106–7, 109
complementary outputs 75
components 1
computation 78
computers
   flip-flops 75
   host 138
   memory 31
   priority encoders 138
   programmable integrated circuits 138
   sound 95
continuous pulses 42
continuous sequencers 110–11, 119
   binary counters 110, 111
   Boolean algebra 111
   clocks 110, 111
   crystal oscillators 111
   logic systems 110, 111
   relaxation oscillators 111
   small 110
   specification 110
   state tables 110–11
   states of output 110–11
   timing diagrams 111

counters *see* binary counters; decimal counters
counting pulses 100–19
   binary counters 100–4, 116
   clocks 105–9, 117
   continuous sequencers 110–11, 119
   one-shot sequencers 112–15, 119
coupling capacitors 83, 85, 124, 127
CPU (central processing unit) 142
crystals 105–6, 111, 113, 160
current
   amplifiers 83, 85, 87–8, 90–1, 93–4
   diodes 7, 25, 27
   direct (d.c.) 83, 90, 124, 127
   drain 10
   light-emitting diodes 7
   logic gates 10, 41
   measurement 2–3
   MOSFETs 10
   op-amps 24
   sink 25
   voltage divider circuits 20, 22
   Zener 27
   see also alternating current (a.c.)
current-voltage characteristics 27, 28

data capture
   flip flops 78
   latches 72, 74, 76
De Morgan's Theorem 56, 57, 58, 61
   NOR gate bistables 70
decimal counters 103, 104
decimal displays 103
   seven-segment 103
decision ladders 151, 153–4
decision-making 147, 151, 153–4
decoders
   address 115, 142, 147, 151
   binary counters 103
delaying signals 29–32, 35, 42
   changing charge 29
   charging 30
   discharging 31
   shifting edges 29
   time delay 32
demultiplexers 54
   circuit symbol 160
   latches 73
   read-only memory 115
digital inputs 1–4, 15
   charge 2
   maximum rating 3
   meters 2
   power 3
   pull-up resistors 4
   resistance 3
   switches 1–2
digital pulses 37–48
   oscillators 42–6, 47–8
   single spikes 37–42
digital signals 1, 19
   combining 5–9, 15–16
   from analogue 19–35
   high 1
   low 1
   see also digital inputs; digital pulses; storing signals
digital systems *see* simple digital systems

digital-to-analogue converters 95, 123
   see also analogue-to-digital converters
diodes 23, 24–8, 33–4
   anodes 25
   cathodes 25
   circuit symbol 160
   clamp 25, 38, 39, 40
   forward biased 25, 26, 39, 44
   reverse biased 25, 26–7, 44
   silicon 25, 26
   Zener 26–8, 160
direct current (d.c.)
   audio systems 124, 127
   blocking 124, 127
   microphones 83, 90
distortion 85
drain current 10
drivers 10–11
   maximum power ratings 11
   MOSFETs 10–11, 46, 110–11, 125, 160
   voltage 122

earth, virtual 91, 94
economies of scale 140
edge signals 37–8, 40–2, 44
   shifting edges 29
   see also falling edges; rising edges
edge-triggering 76, 77, 78
enable inputs
   latches 72–4
   one-shot sequencers 113
   oscillators 43, 44
EOR function 58
EOR gates 9, 59
   block diagrams 12–13
   circuit symbol 160
   flip-flops 78
   microcontrollers 144
   monostable action 42
   truth tables 58
exponential voltage change 30, 31

falling edges 37, 40–2, 44
   audio systems 128
   clocks 106, 108, 109, 111
   continuous sequencers 111
   monostables 149
   one-bit counters 100–1, 104, 109
   one-shot sequencers 112, 113
farads (f) 30, 40
feedback loops 41
   bistables 67, 68
   see also negative feedback
filters, audio 124, 127–33, 135–6
   active 130–2
   analysis 131
   bandpass 133
   bass-cut 129, 132, 133
   break frequency 128, 129, 130–3
   coupling capacitors 127
   gain–frequency graphs 129, 131–3
   impedance 127
   passive 130
   RC network 128–31
   testing 128
   treble-cut filters 129, 130, 133

Index

two lines 131
flat response 123, 124
flip-flops 75–8, 80–1
  binary counters 100–2
  capacitors 76, 77
  circuit symbol 75, 160
  clocks 76–7, 100–1, 105–8, 141
  complementary outputs 75
  computation 78
  D flip-flops 75–8, 100–2, 141, 160
  edge-triggering 76, 77, 78
  four-bit registers 78
  frozen 76, 78
  master and slave 77
  one-shot sequencers 112–14
  quiz referee 76
  registers 78, 141
  set and reset 75, 76, 78
  switch bounce 77
  timing diagram 75, 76
flowcharts 137–8, 145–54
  analysis 146
  branches 147
  chunks (routines) 149–51, 153–4
  decision ladders 151, 153–4
  the end 154
  initialization routines 149
  links 149
  modular approach 149
  output boxes 146
  process boxes 146
  start boxes 146
  stop boxes 154
  symbols 160
  system 149–51
  translation into machine code 138
forbidden states 70
forward bias 7, 25, 26, 39, 44
four-bit counters 101, 113
frequency
  formula 159
  oscillators 45
frequency division 108–9, 113
  by anything 109
  division by three 108–9
  timing diagram 109

gain–frequency graphs 124, 129, 131–3
  see also transfer characteristics
gates 10
guitars 93

hardware 138, 141–4, 155
  analogue inputs 143, 144
  analogue-to-digital converters 141, 143–4
  expanding systems 142
  large systems 142
  memory 142–3
  ports 142
  processing words 143
  registers 141–2, 143
  small systems 142
  systems 141
  two-bit conversion 144
heater circuits 137, 140, 160
heating power 3, 11

hertz (Hz) 45
hexadecimal coding (hex) 145–7, 150
hours 106, 107
howling (audio) 123
hysteresis 43

impedance 121–2, 126–9, 131–3
  matching 121–2
in parallel connections 2
  D flip-flops 141
  op-amps 26
in series arrangements 2
  latches 77
  one-bit counters 101
  resistors 6, 19, 20, 27
input pins 141, 142
input protection 39
inverters 160

jitter 113

keyboards 19
known gain 89–92, 98

lamps 160
latches 72–4, 80
  bistables 73
  bit storage 72
  circuit symbol 160
  data capture 72, 74, 76
  data input 72
  demultiplexers 73
  enable inputs 72–4
  fine detail 73
  four-bit 72
  frozen 72, 73, 77
  inside 73
  master-slave pairs 77
  NAND gate only 74
  in series 77
  set, reset and enable 74
  stopwatches 107
  timing diagrams 72, 74
  transparent 72, 73, 77
  word storage 72
light meters 152–4
light sensors 21, 23–4
light-dependent resistors (LDRs) 21, 154, 160
light-emitting diodes (LEDs) 6–7
  barchart sensors 26
  block diagrams 12–13
  circuit symbols 160
  continuous sequencers 110, 111
  forward biased 7, 26
  indicating low 7
  large NAND only logic systems 61
  LED bias 7
  logic gates 6
  polarity 7
  programmable integrated circuits 140, 146, 152–4
  range sensors 26
  reverse biased 7
  seven segment 160
liquid crystal displays (LCDs) 105
load 10

logarithmic scales 123
logic gates 5–9
   circuit symbols 12, 160
   clamp diodes 39
   CMOS 5, 10, 29, 39–41
   continuous sequencers 111
   delaying signals 29
   demultiplexers 54
   drivers 10
   LED bias 7
   light-emitting diodes 6–7
   logic system design 54
   NOR gates 8, 14, 59, 160
   one-shot sequencers 114–15
   output indicators 6
   polarity 7
   programmable systems 139, 143, 144
   series resistors 6
   switching outputs 10
   threshold voltage 29, 40
   truth tables 5, 8–9, 10, 49
   see also AND gates; EOR gates; NAND gates; NOT gates; OR gates
logic systems 49–65
   analysing 50, 51, 63
   Boolean algebra 51
   decoders 103
   design 54–8
   and demultiplexers 54
   EOR function 58
   and multiplexers 54–5
   NAND function 57
   NOR function 56
   Race Hazard Theorem 56, 57
   Redundancy Theorem 56, 57
   simplifying expressions 55
     large 61–2
     one-shot sequencers 112
     only NAND gates 59–63
     small 50, 51
     system behaviour 50
     truth tables 49–53, 64
loudspeakers 46
   audio systems 120, 123–6
   circuit symbol 160
   peak power 125, 126
   peak voltage 125
   power transfer 126
   in telephones 82
   tone control 123, 124
   treble boost 124
   voltage drop 125, 126
   voltage followers 86, 87

machine code 138
master-slave pairs 77
maximum power rating 3, 11
memory 67, 142–3
   addresses 142
   computer 31
   locations 142
   structure 142
metal oxide semiconductor field effect transistor drivers *see* MOSFET (metal oxide semiconductor field effect transistor) drivers
meters 2

microcontrollers 137–60
   analogue inputs 143, 144
   analogue-to-digital converters 141, 143–4
   circuit symbol 160
   expanding systems 142
   hardware 138, 141–4, 155
   input pins 141, 142
   input registers 142, 145, 148–51
   large systems 142
   memory 142–3
   output pins 141
   output registers 142, 147–54
   pins 141, 142
   ports 142, 145–6
   power supply 141
   processing words 143
   programmable systems 137–40, 155
   reference signals 144
   registers 141–3, 145, 147–54
   reset terminal 141
   small systems 142
   software 145–54, 156–8
   two-bit conversion 144
microphones 82–3
   audio systems 120, 121, 122
   circuit symbol 160
   electret 82, 121
   impedance matching 122
   output impedance 121
   polarity 83
   and summing amplifiers 99
   time constant 83
minutes 106
mixers 96
mobile phones 82
monostables
   action 41–2
   circuit symbol 42, 160
   clocks 109
   continuous pulses 42
   delayed pulses 42
   frequency division 109
   one-shot sequencers 112
   programmable 149–50
   pulse duration 41, 42, 149, 150, 159
   read-only memory 115
   resistors 41
   square wave 43
   trigger 42
MOSFET (metal oxide semiconductor field effect transistor) drivers 10–11
   circuit symbol 160
   continuous sequencers 110–11
   power amplifiers 125
   source 10
   speakers 46
motors 160
mouse 19
multiplexers 53, 143
   address decoders 142
   circuit symbol 53, 160
   logic system design 54–5
multipliers 160

NAND function 57
NAND gate bistables 71

Index

active-low inputs 71
algebraic analysis 71
NAND gate only latches 74
NAND gates 9
  analysis 63
  Boolean expressions for 59
  circuit symbol 160
  clocks 109
  flip-flops 77
  frequency division 109
  large logic systems 61–2
  logic system design 57, 58, 61
  monostable action 41
  one-shot sequencers 114
  only NAND gates 59–63, 64–5
  priority encoders 139
  truth tables 57, 63
  using for AND gates 60
  using for NOT gates 59
  using for OR gates 60
negative feedback 82–99
  amplifiers 82–99, 121, 125–6
  known gain 89–92, 98
  summing signals 93–6, 99
  voltage followers 86–8, 97–8
nibbles 142, 143, 145, 148–50
NOR function 56
NOR gate bistables 69–70
  active-high inputs 69, 73
  algebraic analysis 70
  forbidden states 70
  and latches 73
NOR gates 8, 14, 59, 160
NOT functions 53
NOT gate bistables 67, 68
NOT gates 5–7
  amplifiers 85
  binary counters 101
  Boolean algebra 51, 54
  capacitors 29
  delaying signals 29, 30
  hysteresis 43
  light sensors 23
  logic system design 54, 55, 56
  NAND gates as 41, 59
  one-shot sequencers 113
  oscillators 43–5
  read-only memory 115
  Schmitt trigger 43–4, 102–3, 140, 160
  single spikes 37
  spike-to-pulse conversion 40
  stopwatches 107
  threshold voltage 40, 43, 44
  trip points 44
  truth tables 50, 54, 56

ohms (Ω) 3, 40
one-bit counters 100–1
  clocks 109
  continuous sequencers 111
  in-series 101
  stopwatches 107
one-shot sequencers 112–15, 119
  circuit operation 112–13
  clock terminals 112
  enable input 113

jitter 113
logic systems 114–15
oscillators 112, 113, 114
race hazards 115
read-only memory 115
reset 112, 113
resting state 114
timing diagrams 112, 113
truth tables 114
op-amps (operational amplifiers) 23–8, 33–4, 82, 86–93
  audio systems 121, 125–6, 130, 132
  barchart sensors 26–7
  circuit symbol 160
  inverting inputs 24
  known gain 89–92, 98
  L272 125
  microcontrollers 137, 144
  non-inverting inputs 24
  open-loop gain 87
  output impedance 121, 126
  output saturation 24, 25, 26
  output voltage formula 159
  in parallel 26
  range sensors 26
  saturation levels 87, 89
  short-circuit protection 126
  stable signals 28
  supply rails 24, 90
  TL084 24, 87, 90
  transfer characteristics 87, 88
open-loop gain 87
OR gates 8
  Boolean algebra 51, 54
  circuit symbol 160
  clocks 107
  latches 74
  logic system design 54, 55
  microcontrollers 144
  read-only memory 115
  truth tables 50, 52, 60
  using NAND gates for 60
OR rules 52, 53
oscillation, audio systems 123
oscillators 42–6, 47–8
  circuit symbol 160
  clock inputs 105, 107
  crystal 105–6, 111, 113, 160
  enable input 43, 44
  frequency 45
  hysteresis 43
  mark–space ratio 43
  microcontrollers 141
  one-shot sequencers 112, 113, 114
  oscillated 46
  period of the output signal 43, 45
  relaxation 43–5, 111, 113, 159
  Schmitt trigger NOT gates 43–4
  sound 46
  timebase 45
oscilloscopes 45
  amplifiers 82, 83, 85
  testing audio filters 128
output indicators 6
output pins 141
output signal, period of 43, 45

output switching 10–11
output terminal 1

pins 141, 142
polarity 7
ports 142, 145–6
position sensors 21–2
   bottom terminal 21
   top terminal 21
   wipers 21, 22, 26
potentiometers 21–2, 26
   circuit symbols 160
   linear 22
   mixers 96
   rotary 22
   volume control 125, 130
power 3
   audio systems 120, 122, 125, 126
   formula 159
   loudspeakers 46
   maximum rating 3, 11
   and pull-up resistors 4
   supply 141
   see also heating power
power amplifiers 120, 125
pre-amplifiers 120–4
   voltage gain 123
prefixes 4, 160
premature end 40
printed circuit boards 139
priority encoders 138–9
   three-bit 139
programmable integrated circuits (PICs) 137–44
   analogue-to-digital converters 152–4
   economies of scale 140
   flowcharts 138, 145–54
   hardware 141–4, 155
   light meters 152–4
   PICs count 148
   response time 139
   software 145–54
   test-and-skip loops 148, 150
   truth tables 150–1
   see also microcontrollers
programmable monostables 149–50
programmable systems 137–40, 155
   economies of scale 140
   flowcharts 137–8
   hardware 138
   logic gates 139, 143
   more systems 140
   priority encoders 138–9
   programs 137, 138, 140
   software 138
programs 145–54, 156–8
propagation delay 107
pull-down resistors 6, 87
pull-up resistors 4
pulse duration (width) 40
   monostables 41, 42, 149, 150, 159
pulse tables 100–3, 114
pulsed high inputs 69
pulses 37–8
   continuous 42
   delayed 42
   premature end 40
   spike-to-pulse conversion 40, 41
   see also counting pulses; digital pulses

quiz referee 76

Race Hazard Theorem 56, 57, 62
race hazards 56, 57, 62, 115
range sensors 26
RC networks 38, 40, 42, 43, 44
read-only memory (ROM) 115
Redundancy Theorem 56, 57, 62
reference signals 144
registers 141–2, 143, 147–54
   circuit symbol 160
   eight-bit 142, 143, 145
   flip-flops 78
   four-bit 141, 142
   input 142, 145, 148–51
   loading 153
   output 142, 147–54
resistance 3
   capacitors 30, 31
   diodes 25
   formula 159
   light-dependent resistors 21
   MOSFET drivers 10, 11
   oscillators 43
   position sensors 21–2
   thermistors 19
   voltage divider circuits 20, 22
resistive sensors 19–22, 33
   angle sensors 22
   barchart sensors 26–7
   position sensors 21–2
   resistor ratios 20
   sensor circuits 21
   setting signals 22
   thermistors 19, 26
   voltage divider calculations 20
resistor ladders 26, 144
resistor ratios 20
resistors 1–4
   amplifiers 83, 85, 87, 89–94, 96
   audio systems 126–8, 130–3
   charge 2, 29
   circuit symbol 160
   current 2, 3
   feedback 131–3
   fixed 26
   input 132–3
   light-dependent 21, 154, 160
   light-emitting diodes 6
   maximum rating 3
   microphones 83
   monostable action 41
   one-shot sequencers 113
   oscillators 44, 45
   power 3
   programmable integrated circuits 140, 148
   pull-down 6, 87
   pull-up 4
   resistance 3
   series 6, 19, 20, 27
   speakers 46
   spikes 38, 39

Index

variable 160
reverse bias 7, 25, 26–7, 44
rising edges 37–8, 44, 76–8, 141, 143
   audio systems 128
   clocks 109
   four-bit counters 101
   monostables 149–50
   one-bit counters 100
   one-shot sequencers 112
ROM (read-only memory) 115

Schmitt trigger NOT gates 43–4, 102–3, 140
   circuit symbol 160
   in programmable integrated circuits 140
screen trace 45
seconds (s) 40
   counting 106
   display 106
   length of 105
security systems 154
sensors
   angle 22
   barchart 26–7
   circuits 21
   light 21, 23–4
   position 21–2, 26
   range 26
   resistive 19–22, 33
   thermistors 19, 26
sequential systems 75, 76
series resistors 6, 19, 20, 27
shifting edges 29
short-circuit protection 126
simple digital systems 1–17
   combining signals 5–9, 15–16
   digital inputs 1–4, 15
   switching outputs 10–11, 16
   system diagrams 12–14, 16–17
simplifying expressions 51, 53, 55
sine waves 85, 86
sink current 25
software 145–54, 156–8
   analogue inputs 152–4
   decision-making 147, 151, 153–4
   the end 154
   flowcharts 145–54
   going low 150
   initialization routines 149
   inputs 148, 152–4
   modular approach to 149
   PICs count 148
   programmable monostables 149–50
   pulse duration 150
   registers 145, 147–54
   steady outputs 153
   test-and-skip loops 148, 150
   toggling bits 154
   truth tables 150–1
sound
   oscillators and 46
   see also audio; audio systems
sound waves 82, 83
   peak values 83
   period 83
spike generators 38, 39
spike-to-pulse conversion 40, 41

spikes
   negative 38, 39
   positive 38
   single 37–42, 47
   suppression 39
square waves 43
   binary counters 101, 108, 109
   clocks 105, 108, 109
   testing audio filters 128
states, stable 67, 76
stopwatches 107
   display freeze 107
storing signals 67–81
   bistables 67–71, 79
   flip-flops 75–8, 80–1
   latches 72–4, 80
   sequential systems 75, 76
sum of terms 51, 53, 58
summing amplifiers 93–6, 99
   digital to analogue 95
   formula 94, 95
   mixers 96
supply rails
   amplifiers 91
   audio systems 125
   bottom 1
   and op-amps 24
   top 1
switch bounce 77
switches 1–2
   audio systems 124
   bistables 68, 69
   block diagrams 12–14
   capacitors 31, 32
   circuit symbol 160
   flip-flops 76, 77
   large NAND only logic systems 61
   NOT gates 6
   programmable systems 138, 148
   pull-up resistors 4
   single spikes 37
   stopwatches 107
switching outputs 10–11, 16
system diagrams 12–14, 16–17
   analysis 13
   block diagrams 12–14, 23, 120–1, 124
   circuit diagrams 12

telephones 82
temperature
   monitoring 137
   thermistor resistance 19
terms 53
   sum of 51, 53, 58
test-and-skip loops 148, 150
thermistors 19, 26, 137, 160
three-bit counters 102
threshold voltage 10, 29, 40, 43, 44
time constants 30, 38, 39, 44
   audio systems 128
   formula 159
   microphones 83
time delay 32
   see also delaying signals
timebase 45
timing diagrams 29, 37

AND gates 8
binary counters 101, 103
bistables 68, 69, 70–1
continuous sequencers 111
decimal counters 104
flip-flops 75, 76
frequency division 109
latches 72, 74
one-shot sequencers 112, 113
oscillated oscillators 46
relaxation oscillators 43
spike-to-pulse conversion 40
tone control (audio systems) 120, 123–4
bass boost 124
block diagram 124
flat response 123, 124
logarithmic scales 123
transfer characteristics 123, 124
treble boost 124
traces 45
traffic lights 110–11
transfer characteristics 5
amplifiers 82, 84–5, 87–8, 92
bass boost 124
op-amps 87, 88
oscillators 43–4
treble boost 124
see also gain–frequency graphs
treble boost 124
triangle waves 85
trip points 44
lower 44
upper 44
true digital 23
truth tables 10
AND gates 8
AND rules 52, 53
block diagrams 13–14
and Boolean algebra 51, 53, 54
brackets 51, 53
designing logic systems from 54–7, 64
EOR gates 9, 13
and logic systems 49–53, 54–7, 64
NAND gates 9
NOR gate bistables 69, 70
NOR gates 8, 69
NOT gates 5
one-shot sequencers 114
OR gates 8
OR rules 52, 53
simplifying 51, 53
software 150–1
states 50
system behaviour 50
table to algebra 53
two-bit conversion 144
two-bit counters 101

virtual earth 91, 94
voltage
bias 7
breakdown 26–7
exponential change 30, 31
threshold 10, 29, 40, 43, 44
see also current-voltage characteristics
voltage amplifiers 120, 124
voltage dividers 19
calculations 20
light sensors 23
setting signals 22
stable signals 28
temperature sensing 26
voltage drivers 122
voltage drop 2–3
amplifiers 83
diodes 25, 27
loudspeakers 125, 126
microphones 83
MOSFETs 10
pull-up resistors 4
resistor ratios 20
voltage divider circuits 20, 22
voltage followers 86–8, 97–8, 125, 130
closing the loop 88
difference amplifiers 87
power in 87
power out 86
pull-down resistors 87
voltage gain
amplifiers 84–92, 121–3, 125, 130–3, 159
formulas 91, 159
voltage–time graphs 26, 45
voltmeters 2, 45, 160
volts (V) 3
volume control 120, 124, 130
resistance track 124
wipers 124

waves
sine 85, 86
sound 82, 83
square 43, 101, 105, 108–9, 128
triangle 85
wipers 124

Zener current 27
Zener diodes 26–8
circuit symbol 160
power rating 27
series resistors 27
stable signals 28